'This highly original and radical book addresses the rapidly growing need for an accessible climate pedagogy which represents the different dimensions of the climate-change challenge and can be adapted to a variety of contexts.'
Ted Shepherd, *PhD, Grantham Professor of Climate Science, University of Reading, UK*

'This book elegantly builds on the pedagogic merit of acknowledging the climate system as a teacher. Weaving together stories of different ways of knowing and being, and of the sciences that speak from within disciplinary and inter-disciplinary boundaries, it gives teachers a conceptual tool kit that helps students see beyond facts and make sense of given structures and networks of interactions and relations. The richest harvest is that teachers can use this book to open young minds to pluralism, values and justice, instill competencies to question given disciplinary knowledge, and make informed choices today, when climate change challenges all our current frameworks for making sense of the world.'
Rajeswari S. Raina, *PhD, Professor, School of Humanities and Social Sciences, Shiv Nadar University, India*

'Written from the perspective of a teacher-scholar, this book demonstrates how it is possible – and why it is imperative – for educators from even the most specialized of scientific disciplines to engage in the transformation of education in an era of climate change. Vandana's message is crystal clear: that teaching climate change must transcend disciplines, disrupt power hierarchies, embrace the complexities of Earth systems and social systems, be radically rooted in climate justice, and be delivered through stories that do not shy away from climate emotion. This book is essential reading for any educator seeking radically new pedagogies that nourish the seeds of transformative climate action.'
Christina Kwauk, *PhD, Social Scientist and Policy Analyst, Founder/Director, Kwauk & Associates; Research Director, Unbounded Associates*

'Educators around the world are beginning to address the climate emergency, and the best examples of climate education interrogate how social inequalities shape the contours of this global problem. This book is a valuable contribution to this growing literature, for Vandana Singh develops a model of justice-based climate education that transcends space, time, and the scholarly disciplines while also taking seriously issues of power and injustice. Educators at all levels will encounter valuable ideas to inform their practice while also learning about how to transform broader educational structures toward a more just and humane future.'

Joseph A. Henderson, *PhD, Associate Professor of Social Sciences*, *Paul Smith's College*, *Co-editor of* Teaching Climate Change in the United States

Teaching Climate Change

Teaching Climate Change: Science, Stories, Justice shows educators how climate change can be taught from any disciplinary perspective and in a transdisciplinary way, drawing on examples from the author's own classroom.

The book sets out a radical vision for climate pedagogy, introducing an innovative framework in which the scientific essentials of climate change are scaffolded via three transdisciplinary meta-concepts: Balance/Imbalance, Critical Thresholds and Complex Interconnections. Author Vandana Singh grounds this theory in practice, drawing on examples from her own classroom to provide implementable ideas for educators, and to demonstrate how climate change can be taught from any disciplinary perspective in a transdisciplinary way. The book also explores the barriers to effective climate education at a macro level, focusing on issues such as climate misinformation/misconception, the exclusion of social and ethical concerns and a focus on technofixes. Singh uses this information to identify four key dimensions for an effective climate pedagogy, in which issues of justice are central: scientific-technological, the transdisciplinary, the epistemological and the psychosocial. This approach is broad and flexible enough to be adapted to different classrooms and contexts.

Bridging the social and natural sciences, this book will be an essential resource for all climate change educators practicing in both formal and informal settings, as well as for community climate activists.

Vandana Singh is a professor of Physics and Environment in the Department of Environment, Society and Sustainability at Framingham State University in Massachusetts, USA. She has been working for over a decade on a transdisciplinary, justice-centered pedagogy of climate change at the intersection of science, society and justice, with particular attention to marginalized communities in India and the United States. She was a 2021 Climate Imagination Fellow at the Center for Science and the Imagination at Arizona State University. She facilitates the Education working group for My Climate Risk, a Lighthouse Activity of the World Climate Research Programme focusing on climate science and communities.

Research and Teaching in Environmental Studies

This series brings together international educators and researchers working from a variety of perspectives to explore and present best practice for research and teaching in environmental studies.

Given the urgency of environmental problems, our approach to the research and teaching of environmental studies is crucial. Reflecting on examples of success and failure within the field, this collection showcases authors from a diverse range of environmental disciplines including climate change, environmental communication and sustainable development. Lessons learned from interdisciplinary and transdisciplinary research are presented, as well as teaching and classroom methodology for specific countries and disciplines.

Institutionalizing Interdisciplinarity and Transdisciplinarity
Collaboration across Cultures and Communities
Edited by Bianca Vienni Baptista and Julie Thompson Klein

Interdisciplinary Research on Climate and Energy Decision Making
30 Years of Research on Global Change
Edited by M. Granger Morgan

Transformative Sustainability Education
Reimagining Our Future
Elizabeth A. Lange

Poetry and the Global Climate Crisis
Creative Educational Approaches to Complex Challenges
Edited by Amataritsero Ede, Sandra Lee Kleppe, and Angela Sorby

For more information about this series, please visit: www.routledge.com/Research-and-Teaching-in-Environmental-Studies/book-series/RTES

Teaching Climate Change
Science, Stories, Justice

Vandana Singh

Routledge
Taylor & Francis Group
LONDON AND NEW YORK

from Routledge

Designed cover image: © Getty Images

First published 2024
by Routledge
4 Park Square, Milton Park, Abingdon, Oxon OX14 4RN

and by Routledge
605 Third Avenue, New York, NY 10158

Routledge is an imprint of the Taylor & Francis Group, an informa business

© 2024 Vandana Singh

British Library Cataloguing-in-Publication Data
A catalogue record for this book is available from the British Library

ISBN: 978-1-032-27858-2 (hbk)
ISBN: 978-1-032-27859-9 (pbk)
ISBN: 978-1-003-29444-3 (ebk)

DOI: 10.4324/9781003294443

Typeset in Times New Roman
by KnowledgeWorks Global Ltd.

In loving memory of my father, Priyaranjan Prasad
(1936–2023)

Contents

1 Introduction
The Climate as Teacher

Imagine the sea ice.

You are on the shore of the Alaskan Arctic, looking northward toward the pole. Before you, the smooth snow of the 'beach' gives way to an uneven, glittering field of sea ice (Figure 1.1). Sea ice is frozen sea water, afloat on the liquid ocean; in certain seasons, it extends from the shore toward the North pole, and is thick enough to walk on. You can see the pressure ridges that form when the heaving water underneath pushes chunks of ice against each other. Under the April sun, it is still light at 10 PM, and the temperature is −30°C.

In my undergraduate physics classroom near Boston, Massachusetts, I begin our explorations of the climate crisis through an evocation of place. Up in the high Arctic, sea ice is rapidly diminishing due to climate change, with serious consequences for all who depend upon it – from polar bears to the Indigenous people, the Iñupiat, who have lived in this region for thousands of years. Why should we, the people of the South – because *almost everywhere on Earth* is south of the North Slope of Alaska – why should we care?

Our explorations begin with a story.

The Story of the Scientist and the Elder

In the days when sea ice was still thick in the Springtime, a Scientist went to Utqiagvik with a team of fellow scientists to study the ice. He knew it well: its physics and chemistry, its seasonal cycles, but he wanted to expand his understanding through field measurements. It was a perfect day, as the group walked from the shore over the ice. Clear sky, and nothing in the weather forecast to cause concern. Their footsteps crunched solidly, giving no inkling that beneath the surface layer of ice was frigid liquid water. It was the sea ice thickness, among other properties, that they intended to measure. The ice glittered where the low light of the sun caught the pressure ridges that were formed when the liquid water underneath pushed chunks of ice over each other. They walked to the edge where the sea ice met the liquid ocean, and proceeded to set up camp.

As they were doing so, the Iñupiaq Elder who was their guide suddenly said, without preamble, that they had to get off the ice right now! *They were*

DOI: 10.4324/9781003294443-1

Figure 1.1 A plane approaching land over the sea ice north of Utqiagvik, Alaska.
Photo: Author.

skeptical, because the day was perfect, and there were no indications that anything might go wrong. But the urgency in his voice compelled them to obey. As they walked reluctantly back toward the shore, there was a deafening crack behind them, and the region of sea ice on which they had set up camp broke off and floated into the Arctic ocean.

In those days, with no coastguard presence, being borne off into the harsh Arctic environment at the mercy of unpredictable currents was close to a death sentence. The Elder had saved their lives.

The Scientist had thought there was only one way to know sea ice, and it was with this purpose that he had gone out to its far edge with his instruments. But the Elder knew sea ice in a completely different way. The ice had spoken to him, but not to the Scientist. The Scientist never forgot the experience.

This story is a dramatized version of a true account I heard from a scientist in Alaska. I have been telling this story, or variations thereof, in my general physics classes for the past eight years. Afterwards, I ask my students what questions arise in their minds. Students want to know, more than anything, how the Elder knew the ice was going to break. They also want to know more about sea ice, why it broke, and why it might be a subject worthy of scientific study. Some of them wonder why we are focusing on a remote place like the Arctic.

I promise them that our exploration of climate change will start with their questions. But I have one more place to take them, one more story to tell. Now I ask them to imagine a vast, desertified landscape of low, undulating hills. Unlike the Arctic, this place is warm – it is in the subtropics of India. Once a great forest covered this landscape; now, there is nothing here but emptiness

Figure 1.2 Parvati Devi and fellow villagers, Jharkhand, India.
Photo: Somnath Mukherji.

broken by new roads busy with trucks, and mines, where the acrid dust from stone quarries forms clouds kilometers long. A little further in this devastation, however, is a small remnant of a forest and a village.

Here is a true story that is set in that impoverished village. It was told to me by one of the villagers, a woman called Parvati, speaking on a borrowed phone. A friend who has a long history of engagement with the community provided me with some context. Figure 1.2 shows a photograph of Parvati and her fellow villagers. She is second from left.

The Story of the Village Women of Jharkhand

More than twenty years ago, there used to be thick forests here, alive with wild animals, a variety of birds, bears, even tigers. The summers were cool, and water plentiful. The villages near the forests thrived on a combination of subsistence agriculture and forest produce, including fodder and water. Then, about twenty-two years ago, things started to change.

The great forests were razed down, a process that has accelerated in the past six or seven years. Quarries and mines came up, and broad roadways where trucks plied incessantly, carrying away mineral ore and wood. The water table dropped, and agriculture in the villages started failing. Summers became unbearably hot, and the last remaining forest near the village started to dwindle as the pressure on it from loggers and villagers increased. The wild animals were gone. Malaria was on the rise.

As a marginalized community, with little or no formal education, the villagers had nowhere to turn. So, Parvati and some of the other village women decided to take matters into their own hands. They began by patrolling their remaining forest in the early morning, in groups of three. When they saw loggers or other miscreants, one of the women would run back to the village and return with more people. Confronting loggers was often dangerous. The women also nurtured the forest back to health by limiting use of its resources, and by digging ponds for the animals. They made mud and stone check dams to prevent the water from the streams from going into the desertified surroundings and evaporating.

Over twenty-two years later, the forest has regenerated. The tree-trunks have thickened, and the sound of rushing water is sweet to the ears. The water table has gone up, and agriculture has become productive again. Summers are not as hot, and the malaria menace has diminished. Birds and animals, with the exception of the tiger, have returned. The women continue their work to this day. Asked why she does this work, Parvati says: for the people and for the animals.

When I tell this story to my students, responses include surprise that poor, rural women without formal education could work such a miracle. How did they know what to do to regenerate the forest? Students also ask what this story has to do with climate change. There is no mention of climate change in the story, only of deforestation and its impacts. In this story, climate change is not the reason for the villagers' predicament. Then why tell it?

Before engaging with these questions, or the ones that arise from the Alaskan Arctic story, let us ask a broader one: why teach climate change at all? And why teach it in an undergraduate physics classroom (my context) or, in fact, in any other course from biology to literature, psychology to economics, as I argue in this book? Why not let it stay where it seems to naturally belong: in Earth science courses?

Climate change is arguably the single greatest threat to the well-being of humankind and the biosphere. In Earth's four-and-a-half-billion years of history, climate has changed many times due to entirely natural causes. However, it is no longer in doubt that 'human activity,' particularly the large-scale burning of fossil fuels and the destruction of natural habitats, is the key driver of current climate change. As I will establish in subsequent chapters, the climate crisis is also intimately connected with other social-environmental problems, which it inevitably exacerbates – it is part of a *polycrisis* of interdependent crises. The consequences are dire. With increased global average surface temperature, about 1.1°C at the time of writing in 2023, we are already seeing anomalous patterns such as extended heat waves, large-scale forest fires, changing precipitation, the poleward migration of diseases, the unraveling of ecosystems, and sea level rise. And yet, nations and corporations have not risen to the challenge – every year, the problem gets worse as measured in carbon-dioxide emissions, which continue to increase (IEA 2023). In March, 2023, the UN's Intergovernmental Panel on Climate Change issued a Synthesis report (IPCC 2023) of its Sixth Climate Assessment that warns of increasingly serious consequences, including the likely crossing of the 1.5°C threshold in a mere six years (instead of 2100).

This is one compelling reason for educators in all disciplines to teach about the climate crisis and its implications. But I didn't always begin lessons in climate change by telling stories. When I first decided to teach the subject in all my physics courses, I was motivated by an ethical compulsion – since young people were disproportionately affected by climate change, despite not having lived long enough to contribute significantly to the problem, I felt it my duty as an educator to inform them about it so that they could anticipate the changes ahead, and perhaps become changemakers in an uncertain world. So, I aligned basic climate science with physics topics, introducing, for example, sea level rise as an application of the physics of fluids, and the greenhouse effect understood in the context of the sun's electromagnetic radiation. My first couple of attempts were dismal failures. Not only did the students' cognitive understanding suffer from piecemeal learning of disparate topics scattered over the semester, but also their affective response was disheartening. My naïve presumptions about arming them with knowledge so they could become changemakers soon dissipated when I found, instead, that even a fragmentary understanding of climate science and its impacts were resulting in frustration, despair, anxiety, and apathy.

It was this failure that helped me realize, first, that teaching *just* the science was not enough and could even be counterproductive. This meant that I had to teach myself aspects of the climate problem that were outside science; in other words, I needed to become a student again. A few years later, I found myself on the North Slope of Arctic Alaska, where the sea ice stretched out from the land for a couple of miles. In this frozen landscape, a number of generous people – scientists, scholars, and residents, including Indigenous Iñupiat, helped me understand the multiple implications of melting sea ice. Through their stories, and the work of multiple scholars, I began to realize that I needed a radical re-orientation: to think of the *climate system as teacher*.

Education ought to be a powerful ally of climate mitigation and adaptation. Yet, as scholars have established, mainstream education has failed us (Kwauk 2020). Its failure mirrors the failure of nations and corporations and the UN to take meaningful and effective action. As I will establish in this book, among the many complex reasons for these failures to act is this: *climate change challenges all our current frameworks for making sense of the world.* It therefore escapes and confounds conventional ways of thinking, planning, and acting. My Alaskan experience taught me to approach the climate problem with requisite humility, abandoning, to the extent possible, my preconceived frameworks, and acknowledging the phenomenon itself as teacher. Through the diverse voices of local and Indigenous peoples, scientists and scholars, and through immersive (albeit brief) experiences, I have tried to listen to the climate problem to see what it can teach us.

The two stories I've shared, when put in context, contain certain key lessons about the climate. One, the climate problem spans *vast scales of space and time*. The climate crisis is a global phenomenon, but we experience it in locales or climes. It manifests differently in Northern Alaska than in Jharkhand, but the local and the planetary are both important. Temporally, it includes both the slow rise of Earth's average global surface temperature since the time of the industrial revolution, and the dramatic annual shrinking of Arctic sea ice, or the suddenness of a storm made more likely by climate change, or a heat wave in Jharkhand. Two, the climate problem is *inherently transdisciplinary*. How communities like the Iñupiat of Alaska or the Jharkhand villagers experience climatic changes depends on the distinct histories of colonialism and socio-environmental changes due to the economic imperatives of modernity and power. Their relationship with the land emerges from examining questions like 'How did the Elder know the ice was going to break?' or 'How did the village women know how to regenerate a forest?' which lead directly into the disciplines of sociology, anthropology, and environmental history. But an additional lesson from the Jharkhand story is this: that the predicament of the villagers was brought about by deforestation, not climate change. Jharkhand is one of the most climate-vulnerable regions in the world, so the climate problem enters the story through its broader context. The Jharkhand story teaches us that there are other large social-environmental problems than climate change. A later exploration will help us to understand that all of these problems are related at the root. Thus, climate change is *not* an isolated problem, either in a disciplinary sense (as a purely scientific-technological issue) or in the sense of being unrelated to other large problems; it is a symptom, not the disease.

The justice dimension of the climate issue is evident in these two stories. Neither the Iñupiat of Alaska nor the Jharkhand villagers are responsible for creating the climate problem. They have been at the receiving end of multiple oppressions through history. Yet – along with other marginalized groups, including the young – they are the ones disproportionately affected by the climate crisis. Often, the groups that are the most marginalized have creativity and agency, new visions to offer. However, they are invisible to most people, including policymakers. The IPCC 2018 report goes as far as saying that climate change cannot be solved without equity. When we ponder why the climate crisis has not met with a meaningful response from nations and corporations, given how long scientists have been warning us about it, the question of power hierarchies comes to mind: power hierarchies as drivers of the crisis that are also preventing meaningful change. There are many aspects of justice and power that become apparent in our explorations, including multispecies justice. Thus, the climate problem is *inherently a problem of power, justice, and equity.*

In the classroom, I want students to find connections between our own lives and landscapes to those of the communities in these stories. As part of the wider context-setting, we visit our planetarium to look at satellite images of the Earth – an exquisite blue-green sphere, with swirls of clouds and the white polar ice caps. Rotating the image, we can see both the Alaskan North Slope of the Arctic and the rapidly desertifying state of Jharkhand in India.

There are no boundary lines of nation and state in this view – in fact, there are no obvious signs of human habitation. The Earth looks like something dynamic, alive, whole. It is hard to imagine that what happens in one place does not, somehow, affect another. But even within the local landscapes of the stories, an investigation of the questions: *how did the Elder know the ice was going to break? how did the village women know how to regenerate a forest?* reveals the complex relationships between and within human and biophysical systems. (Here, I use 'complexity' in a special sense that I will elaborate in Chapter 4.) The sea ice is central to Iñupiaq cultural identity and survival through its role in the Arctic ecosystem, all the way from ice algae that form beneath the sea ice to the polar bears and seals that walk on it, and the bowhead whales that swim underneath and beyond it. Similarly, regenerating a forest requires a recognition of the interdependence of the various species of animals and plants, and an understanding of the connection between forests and the hydrological cycle. Thus, humans are among multiple diverse actors in these social-environmental systems, and a practical knowledge of the dynamic interrelationships between them is crucial. This systems view is reinforced by the images of Earth from space. With a little prompting, students are able to discover for themselves the five natural subsystems of the Earth as defined by Earth scientists: the land (lithosphere), air (atmosphere), water (Hydrosphere), ice (cryosphere), and life (biosphere). But instead of distinct components of a machine, these subsystems appear entangled and interrelated, giving Earth the look of a blue-green pearl, swimming in space. At this early stage in the semester, we simply pose questions such as: how do different subsystems affect each other? Or more specifically, does Arctic climate change affect Jharkhand? Does deforestation in Jharkhand affect Arctic climate change? What about climate change here, where we live and work? All of these questions have to do with recognizing and exploring the Earth's social-environmental systems as complex systems.

The two stories I have told above are centrally about change. Climatic and environmental changes can potentially cross points of no return (on certain timescales), where the system changes to another state entirely. If sufficient areas of forest cover are destroyed in Jharkhand, the ecosystem will undergo changes that are likely irreversible on human timescales. If Arctic sea ice vanishes, the coastal Arctic ecosystem will cross a threshold into a new state. The presence of critical thresholds – limits and boundaries – is a manifestation of the inherent complexity crucial to understanding the climate problem. At the global scale, the new work on planetary or Earth system boundaries points to the possibility of tipping points causing irreversible changes in the entire Earth system. A very important emerging area of research is concerned with how changes in local and regional phenomena might result in a cascading, domino effect that would cause larger scale Earth system shifts. What this means for human societies, which have their own complex dynamics and thresholds of change, is an open question.

Thus, climate change (and related phenomena) exhibits *complex interconnections and relationality*, which consequences for impacts and actions.

Of course, I don't rely solely on the two stories I have told above; I have a suite of carefully curated stories from scholarly work, direct experience, news reports, and speculative fiction that are invaluable as teaching tools. Well-chosen stories, whether real-life narratives or fictive tellings, can be powerful vehicles for moving across disciplinary boundaries, for revealing complexities, and for amplifying unheard voices. As scholars have recently noted, stories are also ontological tools, and stories that counter dominant narratives are crucial for the onto-epistemological shifts that are needed if we are to engage meaningfully with the climate problem. In addition, certain kinds of stories allow the non-human actors to take center stage, whether these are other animals, such as polar bears or bowhead whales, or aspects of weather, climate, and geography, such as the sea ice. As I will elaborate in Chapter 5, storifying climate science makes the science accessible and meaningful to non-scientists, expands the imagination, and dissolves the subject–object barrier toward a more 'participant-observer' role. Interestingly, a recent epistemological expansion within climate science that embraces the power of narrative – the idea of a physical climate storyline, described briefly in Chapter 4 – has emerged as a promising way to make climate science meaningful to people and communities on the ground.

Before I elaborate on the key ideas in this book, I want to take a moment to introduce and situate myself, acknowledging the impossibility of complete objectivity – we bring our personal experiences, biases, and proclivities to every task, and therefore I believe that being self-aware, transparent, and honest is paramount. Briefly, I am a particle physicist by training, born and raised in India, where I have a number of personal and work-related connections. I acquired an early interest in the environment and environmental justice as a teen growing up in New Delhi, as a member of the still extant environmental justice action group Kalpavriksh. A seminal experience in my life was a trip with Kalpavriksh members to the Himalayas at the invitation of the now famous Chipko movement, where I experienced for the first time the lives, struggles, agency, and creativity of the rural poor – however, it took years for me to make sense of that experience, which continues to reverberate in my life. In my twenties, I went to the United States to obtain my Ph.D. in theoretical particle physics, studying properties of sub-nuclear particles called quarks. Following a postdoctoral fellowship in India, I had an inadvertent ten-year break from academia. I then returned to a small and lively public university in the Eastern United States, where I am now a professor of physics and environment. Here, I rediscovered my love for teaching, which I had first experienced as a teaching assistant in graduate school. Many of my students are first-generation college goers, and they are increasingly diverse in terms of race and ethnicity, including a number of recent immigrants from Latin America and Africa. Most of them work extremely hard at menial jobs to pay for college, sometimes as much as 25 hours a week. In a fixed-mindset educational system that labels students as gifted, average, etc., most of my students have never been challenged intellectually in ways that can unlock their potential. It is my privilege and immense pleasure to show to them that they can expand their horizons, broaden and deepen their understanding, and excel in their academics beyond what they had ever imagined.

I became deeply concerned about climate change in 2007, when my university held a week-long across-disciplines climate teach-in. I taught myself climate science basics from textbooks, beginning with the concise and elegant *Global Warming: Understanding the Forecast* by David Archer, and started teaching the basic science from 2010 onward. It is through my students – their intelligence, their misconceptions, their struggles, their hard work, and their creativity – that I learned how to teach climate change better. It is to them, therefore, that I owe the largest debt.

I am also a writer of literary speculative fiction – an umbrella term that includes science fiction and fantasy. In science fiction in particular, the story takes place against the canvas of the universe itself; as in the ancient epics of multiple cultures, the relationship between the human and the non-human biophysical universe is paramount (a relief from the exclusively human-centric, solipsistic obsessions of much of Anglophone mainstream literary fiction). Researching for a speculative fiction story can involve – for example – such diverse subjects as methane outgassing on the Arctic seabed, intergenerational trauma, the place of the cosmos in a particular culture, the phases of Jupiter as seen from one of its moons, and the possible societal responses to a paradigm shift, sometimes all for a single story. Thus, transdisciplinarity has always felt natural to me, and the power of narrative, evident.

As a person born and raised in India, I am well aware of the legacy of colonialism and its long shadow (including the irony of writing in English), and the ways in which inequality and power manifest in our lives. I am a woman under most lights, a 'person of color' in America, and a caste-privileged person in India. In my family, education and learning across disciplines were greatly valued, and a history of resistance to British rule for some members of my grandparents' generation helped inculcate a healthy disrespect for authoritarianism of all kinds. Growing up in Delhi alongside wild creatures that co-existed in the interstices of a human-made world, I became aware early on of the worlds of the nonhuman, and that most people were uninterested or unaware of these. I have been, in many senses, an 'in-between' person for much of my life, including forays from my home discipline of physics across borders into other intellectual realms. I resonate with what the scholar and film-maker Trinh T. Minh-ha writes (Trinh T. Minh-ha 2015) in another context (italics mine):

> The moment the insider steps out from the inside she's no longer a mere insider. She necessarily looks in from the outside while also looking out from the inside. Not quite the same, not quite the other, she stands in

that undetermined threshold place where she constantly drifts in and out. Undercutting the inside/outside opposition, her intervention is necessarily that of both not quite an insider and not quite an outsider. She is, in other words, this inappropriate other or same who moves about with always at least two gestures: that of affirming 'I am like you' while persisting in her difference and that of reminding 'I am different' while unsettling every definition of otherness arrived at.

The broader relevance – and advantages – of this shifting, uncertain positionality will become evident in later chapters.

Finally, I want to make clear what I am not: I am not a climate scientist, nor am I a scholar of education. As should be evident from the Acknowledgments, I have learned and continue to learn from climate scientists and specialists in education, among other scholars. The experiences that have brought me to the point of writing this book include 20 years of teaching undergraduate physics. Key moments in my learning journey include my university's climate teach-in in 2007, my trip to Alaska in 2014, a Black Lives Matter teach-in at my university in 2016 for which I introduced climate justice in a physics class for science majors, a 2016 workshop on interdisciplinarity in STEM organized by the National Academies of the United States in which I was a panelist, a week-long interdisciplinary climate workshop that I co-conceptualized and co-ran at my university for high school science teachers in 2017, and a number of talks presented and heard, as well as discussions and mutual learnings from other concerned educators in India and the United States.

This book is built upon the lessons of the climate problem as I have learned them so far in my ongoing journey. We will see, later, that this pedagogical approach will take us to considerations beyond science, and indeed, beyond climate itself. My attempt in this book is to convince readers from all disciplines and contexts that any effective pedagogy of climate change should, at the very least, embrace its key teachings: the spanning of large spatio-temporal scales, inherent transdisciplinarity, complex interconnections between and within human and biophysical systems, and between climate and other problems, and the centrality of issues of power, justice, and equity. As I will show, no matter the disciplinary background, it is possible to understand and explain key ideas that are outside the educator's discipline. It is my hope that a teacher of psychology will find it possible to integrate important climate science ideas into their curriculum, and a teacher of chemistry will be able to introduce the intersection of justice with considerations of atmospheric chemistry.

In the following chapter, *What is an Effective Pedagogy of Climate Change?* I begin by considering why our current institutions – education included – have failed us. I contrast the key aspects of what might be called the dominant paradigm of modern industrial civilization with the key features or 'teachings' of the climate problem as elucidated above. I then relate this mismatch to the failures of the educational system in rising to the challenge of climate change, and do a brief literature review of the scholarly work in this area. I point out the barriers to effective climate education at the macro level, both globally and in the United States, and how these are made manifest in the microcosm of my classroom. This discussion includes the impact of climate misinformation, cognitive misconceptions that I have encountered that are also mentioned in the literature, as well as the psychological reactions to learning about climate change, some of which can be barriers to action. From this discussion emerge four criteria for an effective pedagogy of climate change: the scientific-technological, the transdisciplinary, the epistemological, and the psychosocial action dimension. After an elucidation of each, I discuss the importance of best practices inspired by the work of Carol Dweck (Mindsets), Ken Bain (Natural Critical Learning Environment), and multiple scholars of transformational learning theory. I point out (and critique) the importance of the epistemic shift that is a central aspect of the latter, and apply insights from the work of physicist and feminist philosopher Karen Barad. In addition, I discuss the need for climate education across the curriculum, and, ideally, a wholesale restructuring of the education system to serve society in the context of major social-environmental problems. As we work toward long-term systemic changes, I suggest that my framework can provide a pathway for educators to start in any particular discipline, and then go beyond it to a transdisciplinary, justice-centered approach that shortchanges neither the natural nor the social sciences.

In Chapter 3, *Science but Not Just the Science: The Whys and Wherefores of a Transdisciplinary Approach*, I elaborate upon the points raised at the end of the previous chapter. I define multidisciplinarity, interdisciplinarity, and transdisciplinarity. I make the argument for a transdisciplinary understanding of climate change, and the problems with a strictly scientific-technological approach. Why is 'just the science' not enough, and perhaps even harmful? After engaging with this question, I consider how we might, in a compartmentalized education system, start from a particular discipline and go beyond it to transcend disciplinary boundaries. Case-based and story-based methods are particularly effective in bringing out the transdisciplinarity inherent in the climate problem, but it is also possible to 'braid' science, social science, and justice when discussing key climate science concepts, as elaborated in a later chapter. I then lay out the preliminary planning process through which an educator might use the criteria for an effective pedagogy of climate change to introduce climate change into their course. To prevent piecemeal learning (a real danger in a course not devoted exclusively to climate change, where climate-related topics might be scattered through a semester), I introduce a unifying visual tool, which is evoked and employed as a means of layering and deepening understanding, each time a climate topic is examined.

Chapter 4, *Science and More than Science: Three Transdisciplinary Meta-Concepts*, is the heart of this book. Beginning with preliminary ideas that include the definitions of climate and weather, and the difference between the two, this chapter lays out the transdisciplinary meta-conceptual framework that introduces essential aspects of climate science while braiding in key concerns and illuminations from the social sciences, including considerations of justice. My contention is that the essential science of climate change cannot and should not be taught as a list of disconnected facts, but through a sense-making conceptual structure, which embraces but ultimately goes beyond the science. The first Meta-Concept is *Balance/Imbalance*. Beginning with a simple interpretation of passive balance, such as a person standing on the toes of one foot, I introduce dynamic balance through a game that educators and students can play, especially accessible and effective for non-science majors and disciplines. Thus, we can introduce and discuss the anthropogenic impact on the carbon cycle, the greenhouse effect and Earth's energy balance. The question now arises: why is this happening? Who is responsible? This is where considerations of economics (endless growth), consumption, increasing urbanization, and social inequality between and within nations can be introduced. The idea of *critical thresholds at local and planetary scales* – including Earth system (Planetary) boundaries – emerges naturally when we consider how systems go from Balance to Imbalance by crossing thresholds that may be abrupt or gradual. Similarly, the meta-concept of *complex interconnections* arises in the context of questions about relationships, feedback loops and tipping points in natural and social systems. Each of these is elaborated in a critical and integrative manner, and the issue of social inequality and justice is threaded through each. I illustrate how the scientific-technological dimension, the transdisciplinary dimension, and the epistemological dimension are realized through these meta-concepts, along with the connective tissue of justice.

Chapter 5, *The Power of Stories: Foregrounding Justice in the (Science) Classroom*, is especially useful for science teachers, but has wide applicability to all disciplines. While questions of justice are threaded through all three meta-concepts, it is from stories, including case studies, news reports, and fiction, that justice issues become real to students. In this chapter, I focus on stories of different kinds that I have used in the classroom. Real-life news reports, case studies, and selected fiction allow for a juxtaposition of race, history (climatic and racial), politics, science, economics, tragedy, and resilience that makes climate change feel real and immediate. I employ ideas from scholarly work on narrative to support my contention that stories are crucial pedagogical tools. I also introduce stories that emerge from science, and useful ways of 'storifying' scientific processes and concepts through enactments and embodied learning. I mention the recent introduction of physical climate storylines in climate science. I discuss briefly different dimensions of climate justice, and the necessity of highlighting the role of colonialism in our current predicament. I then introduce the physical phenomenon of diffraction and its usefulness as a metaphor and a trans-methodology in diffractive analysis.

My work on an educational case study for teaching Arctic climate change taught me the importance of the remote Arctic to global climate. The cryosphere – collectively, those parts of our planet that are covered with snow and ice – is the focus of Chapter 6: *On Thin Ice: Applying the Framework to the Cryosphere.* I begin with an exploration of the Arctic and the role of sea ice in climatic changes from the last million years to the present. Current work on tipping elements in the climate system link the Arctic sea ice and Greenland ice sheet with the circulation of ocean currents, linked in turn with the Amazon rainforest's viability. I integrate the human and natural dimension of climate change in these regions with the geophysical. I bring out all these points through a detailed description of the application of this framework to a transdisciplinary freshman seminar on Arctic climate change.

Chapter 7, *Critical and Ethical Thinking on Climate Solutions*, develops the fourth dimension of the transdisciplinary framework, the psychosocial action dimension. In an effective pedagogy, attention must be paid to the psychological impact of learning about climate throughout the semester. I suggest how the community of the classroom can support each member through acknowledgment of emotions, and also through meaningful community action. Further, the transdisciplinary framework leads naturally to criteria for distinguishing between real and false solutions, an issue of grave importance today when co-optation of the climate problem and massive greenwashing by existing power structures is very real. I show how students can be guided toward co-creating these criteria on the basis of both efficacy and justice, leading to critical evaluation of purported climate solutions, and a discussion of appropriate actions at the local and global scales. In this chapter, I discuss both what students can do in their communities through projects, with some examples from my own experience, and how I teach the problem of connecting local and global in this way.

In Chapters 8 and 9, *Insights from Other Educators: Reimagining Formal Spaces*, and *Insights from Other Educators: Climate Education Outside the walls*, I describe conversations with seven selected teachers and teacher-scholars about their own experiences of teaching climate change, and their comments (for those who could spare their time) on my pedagogical framework. These teachers are not a representative sample by any means; I located them through various networks and chose them on the basis of interesting work in interesting contexts. Most are from the Global South – this is my small attempt to redress the imbalance in research and ideas from the Global South on climate education. Four of these are working in formal educational settings (the subject of Chapter 8) while the remaining three make bridges from formal spaces to the world at large. There is representation from disciplines as distinct as chemistry and anthropology, and countries as far apart as Costa Rica, Zimbabwe, and India. What unites these educators is passion and concern about the climate problem and the polycrisis. Several of them found the time to comment on the possibilities and shortcomings of the pedagogical framework that is the subject of this book.

This last point deserves elaboration. The fact is that effective, meaningful climate education is a new field. There is also context-dependence: what is meaningful and successful in one context may fail in another. Thus, my pedagogy, although designed as a broad enough framework to allow for adaptation and flexibility in different contexts, is certainly not the last word on the subject. My always-evolving approach can make no claim to universality of application. On the contrary, my brief interviews with a small sample of educators indicate that a rich, creative, and generally unacknowledged tapestry of approaches is developing in different regions and situations around the world, especially in the Global South. It is time for the work on the margins to begin to shake the establishment thinking on climate education. Because, as I hope to show in subsequent chapters, the climate problem and its attendant ills confound our existing, dominant frameworks, our pedagogies should be at least as radical as the problem itself. I discuss these points in the final chapter, Chapter 10, *Reflection-Diffraction: The End and the Beginning*. I first elaborate on why the metaphor of diffraction is better suited to a radical, transdisciplinary climate pedagogy. I discuss, from a more conventional perspective, the ways in which my pedagogical framework holds promise, and where the gaps might be. Following this, I illustrate how a diffractive reading of this pedagogical framework – itself inspired by the metaphor of diffraction – gives us something more – possible pathways and new directions, including more radical and relational approaches to our entanglements with the polycrisis.

It is my fervent hope that fellow educators around the world who are not already creating new pathways to change in their classrooms will be provoked, inspired, and encouraged to do so by my experiences in transdisciplinarity and transformative climate education. Our students, future generations, and our planet deserve no less.

References

IEA. 2023. "CO$_2$ Emissions in 2022." Paris. https://www.iea.org/reports/co2-emissions-in-2022.

IPCC. 2023. "Synthesis Report of the IPCC Sixth Assessment Report AR6: Summary for Policymakers." https://report.ipcc.ch/ar6syr/pdf/IPCC_AR6_SYR_SPM.pdf.

Kwauk, Christina. 2020. "Roadblocks to Quality Education in a Time of Climate Change." *Brookings* (blog). February 25, 2020. https://www.brookings.edu/research/roadblocks-to-quality-education-in-a-time-of-climate-change/.

Minh-ha, Trinh T. 2015. "Not you/Like You: Post-Colonial Women and the Interlocking Questions of Identity and Difference." *AGRESTE Magazine*, 2015. https://cultural-studies.ucsc.edu/inscriptions/volume-34/trinh-t-minh-ha/.

2 What Is an Effective Pedagogy of Climate Change?

2.1 A Cognitive and Affective Failure

When I first started teaching the basic science of climate change in my general physics undergraduate classes for non-science majors, my hope was that a fundamental understanding of the problem would warn students of the implications of the crisis, empower them to act wisely and intelligently, and inoculate them against the misconceptions that were (and still are) rampant among the general public in the United States. These naïve expectations were shattered when my first attempts turned out to be failures. Students did not leave the course with a sound understanding of climate fundamentals – instead, in part because climate science topics were scattered throughout the semester wherever they aligned with physics topics, students had a fragmentary and piecemeal sense of the problem that did not address the misconceptions with which they had come into the course. For example, confusion between the ozone problem and climate change, a popular misconception, remained in the minds of some students. I recall one student saying, half in jest, that perhaps we shouldn't be breathing out, since carbon dioxide is a by-product of respiration. The greenhouse effect presented challenges too, including a failure to distinguish between outgoing electromagnetic radiation and carbon dioxide molecules. This cognitive failure of my approach was matched by an affective failure. Rather than feeling empowered to act, several students reported that they felt depressed, helpless, resigned, angry, despairing, and apathetic. A common reaction was: 'I'm just one person, what can I do?' Others took refuge in the belief that something or someone else would 'fix the problem' – world leaders, scientists, new technology. Some typical reactions are noted in Box 2.1.

This experience made me realize also that learning about climate change was an affective burden on students who were already stressed by various factors: educational debt, having to work long hours at tedious service jobs to pay for college, family issues, including responsibility for siblings or a parent, adjustment problems for recent immigrants from Latin America and Africa, psychosocial barriers for first-generation students who were unsure

DOI: 10.4324/9781003294443-2

Box 2.1   Typical student reactions to learning about climate change, from my early pedagogical experiments

- We're screwed
- This is really depressing
- What can be done?
- I'm just one person – I can't do anything about it
- It's not affecting me so I don't care
- There's no hope. Unless we can go to another planet, humanity is doomed
- I never talk about it outside class
- I don't like thinking about it
- Technology will save us
- It's because of the ozone layer
- The more people breathe out, the more they're contributing to climate change
- Why can't we wait until Boston is like North Carolina and then fix it?
- I heard that scientists are in it for the money so how can we trust what they say?

of how to navigate the college experience, to name a few. Recent studies indicate increasing anxiety and depression among US college students (Lipson et al. 2022); additionally, climate anxiety among the young is growing (Tosin Thompson 2021) and learning about climate change can worsen that anxiety (Verlie, Blanche 2021). Some have suggested that learning about the unfolding horror of the climate crisis is emotionally traumatic (Wysham, Daphne 2012) with its attendant feelings of denial, anger, grief, and apathy. While some emotions might trigger interest and desire for action, others could potentially deter meaningful climate action. Clearly, the emotional aspect of learning cannot be ignored.

It became increasingly clear to me that teaching 'just the science' in a fragmentary way that distributed climate science topics through the semester was not only not working, but was counter-productive. Despair, apathy, helplessness, fear, and grief are logical reactions to the seriousness of the climate crisis, but they can also paralyze us into inaction. To deal with these, I would have to go beyond the greenhouse effect and the properties of the carbon dioxide molecule. I would have to cross disciplinary boundaries and become a student again.

2.2 The Climate as Teacher

My journey of learning took me a few years later (2014) to the Alaskan Arctic. As I stood on the frozen shore near Utqiagvik, with the glittering plain of sea ice stretched before me, I had a visceral sense of displacement and decentering, which, along with multiple conversations with generous residents, would eventually result in a radical reorientation to consider *the climate as teacher*. In Alaska, I interviewed scientists, scholars, and community members, including Iñupiaq town officials and a whaler, and following that, read the research on Arctic climate change as well as scholarly work of anthropologists, economists, sociologists, historians, and others. It became clear to me that 'just the science' was hopelessly inadequate in understanding and conveying the reality and implications of the climate problem. For one thing, climate change was centrally a problem of justice. The Iñupiat of Northern Alaska had not contributed significantly to the climate problem, yet, they were being disproportionately affected by it (Huntington et al. 2022). This applied just as much to seals and polar bears in the region, upon whom the Iñupiat depended for survival, especially during the harsh winters. There was also a complex relationship between their current predicament and the history of colonialism and racism in Alaska. Economics played a crucial role, as the towns of the North Slope derived income and employment from leasing their land to oil companies. There was division within the community on whether to support or oppose offshore drilling. There were ecological shifts and changes in the sea ice that the Iñupiat had not seen in over 4,000 years of ancestral memory. These considerations went well beyond the physics and climatological role of sea ice. Therefore, it became clear to me that interdisciplinarity was key for understanding, conceptualizing, and teaching climate change. This was going to be a major challenge in an education system that divided knowledge into disciplines, each in its own watertight compartment. The following year, through a program award from the Association of American Colleges and Universities' STIRS Initiative (Scientific Thinking and Integrative Reasoning Skills), I created an interdisciplinary case study for undergraduate education focused on Arctic climate change and the complex dilemmas of continued oil exploration, collapsing ecosystems and Indigenous cultural survival. However, it took further exploration of community experiences from other parts of the world, notably, India, where I am from, to drive home the lessons of the climate that I have mentioned in Chapter 1.

Among other things, I learned about the need to go beyond interdisciplinarity to transdisciplinarity (more on the distinction in the next chapter), and the crucial role that stories play in order to effect that change. I also learned about the importance of acknowledging the centrality of power – the fact that we live in unequal societies with pyramidal power structures that drive the crisis and pose a barrier to meaningful change. One of the key questions that arose in my mind was this: what kind of scientific framework is necessary to promote a holistic understanding of the basics of climate science? Because non-science majors, 'ordinary' citizens, and non-experts also need to know, and have the right to know what is happening to our planet. Such a framework would not be a disconnected list of facts, but would provide a conceptual scaffolding for people to emplace climate information in a way that would inoculate against misinformation and result in intelligent and ethical ideas for action. Crucially, within the classroom, such a framework would avoid piecemeal learning (which is no learning at all) despite climate topics being scattered throughout the semester. Thus, this framework could potentially be of use to *any* educator in a conventional education system who was able and willing enough to venture outside their discipline, and who wished to teach climate change in their classes from their disciplinary perspective. In Chapter 4, I will share this framework and its grounding in three large trans-disciplinary 'meta-concepts' that have been useful in my classes.

If climate change is inherently transdisciplinary, spans large spatial and temporal scales, is rife with complex interconnections, and is centrally a prob-lem of justice (the key lessons demonstrated by the stories in Chapter 1), then it is perhaps not hard to see why the problem escapes us. Modern industrial globalized societies tend to emphasize short-term thinking (consider election cycles, quarterly and annual reports, and the ubiquity of deadlines in our lives), which preclude attention to history and historical patterns, and equally, attention to the possibilities of the future of the planet. Thus, we seem to live on a linear, disconnected, fragmented time axis. We are spatially disconnected from regions outside the ever-expanding sphere of modern civilization: the Arctic, the Amazon rainforest, the remote Himalayas; therefore these are also absent from our concerns, leading to a fragmented sense of space. This bro-ken, chopped-up experience of space and time in modern industrial cultures is also seen in the way our empathies, attention, and consideration are limited to familial and affinity groups and to the exclusively human. Linear think-ing (specifically, simple linear causality) and lack of appreciation for detail, nuance, and context can often result in 'black-and-white' oversimplifications about how the world works, and denies complex, nonlinear interconnections within and between human and biophysical systems. We live in highly unequal social hierarchies where the power structures promote an acceptance of the status quo, including justifications for the ascendance of the powerful. In hier-archical societies, the education system inevitably reproduces and reinforces the power pyramid, including their replication within the classroom. It denies and prevents transdisciplinarity by barricading disciplines in silos. Complex systems are rarely taught in high school or college, except for specialized graduate and undergraduate courses, and 'teaching to the test' among other factors has undermined free and critical exploration of ideas necessary for cre-ative and nuanced thinking. Any regard for the continuity and scale of space and time – for example in history and geography classes – is undermined by educational approaches that separate the student from their learning and disciplines from one another. My description of what might be considered key features of modern industrial civilization (and their mirroring in education) is, of course, an oversimplification, which conceals contradictions, reforms, and pushback, but admitted as a broad-brush generalization, it is pedagogically useful. When we contrast these features with the key characteristics of the climate problem, it becomes clear that there is a fatal mismatch, and we can begin to understand one reason why climate change has proven to be so intractable.

From these considerations, it becomes evident that the frameworks or paradigms with which modern industrial cultures construct their realities are woefully inadequate when we are confronted with a problem like climate change. It therefore behooves us to become aware of and discard – to the extent possible – our current frameworks when we consider the problem; that is, it is necessary to go to the problem with a childlike openness and humility that acknowledges the need to learn from the problem itself. It is in this sense that I use the phrase *climate as teacher.*

Therefore, let us acknowledge that we are up against a near-impossible task with regard to climate education. It is not just a matter of including certain topics in our curricula; it is also a matter of a fundamental structural and epistemological transformation in our educational system that in turn transforms and is transformed by the social-ecological context. There are multiple barriers to this necessary transformation.

2.3 Barriers to Meaningful Climate Education

Let us elucidate these in greater detail. At the macro level, scholars have identified five major roadblocks to 'quality education in the time of climate change' (Kwauk, C. 2020): low priority for ecoliteracy, lack of a radical vision for education, a problem of definition and scope in education for sustainable development (including a narrow focus on the science, and an approach counter to the principles of transformational learning), monitoring and accountability mechanisms geared to passive 'progress,' and finally, lack of systemic support for teachers to become 'change agents for sustainability.' These also manifest in the microcosm of the classroom, although they do not map onto each other one-to-one. Based on my experience as well as a (non-exhaustive) survey of the literature, I have identified the following inter-related barriers.

1 Knowledge pollution and ignorance: A Gallup World Poll conducted in 2007–2008 (Lee et al. 2015a) on nationally representative samples in 119 countries revealed that 40% of adults around the world had never heard of climate change; further, that number was more than 65% for certain countries, such as Egypt, Bangladesh, and India. More recently, an international survey of 108,946 people in 110 countries, territories, and geographic groups (Leiserowitz, A. et al. 2022) – limited, however, to Facebook users

and therefore not representative – reported on knowledge about climate change, among related matters. In 46 of these regions, many in the Global North, more than 50% of respondents reported at least a moderate amount of knowledge of climate change; however, substantial proportions of respondents in countries of the Global South stated they had never heard of climate change. These two reports, about 15 years apart, both indicate, despite the difference in sampling, a problematic North-South divide with regard to basic awareness about climate change. This is clearly a serious matter since regions of the Global South are more vulnerable to the climate crisis; it is likely that this lack also manifests in the education system.

Beyond basic awareness, a comprehensive understanding of fundamental climate change science, evidence, and impacts is necessary (but not sufficient) for wise climate action. Unfortunately, research indicates a significant conceptual confusion and ignorance among the US public and the world at large about anthropogenic climate change (Leiserowitz, A., Smith, N., and Marlon, J.R. 2010; Lee et al. 2015b), leading to, for example, confusion between the climate crisis and the ozone problem ('the hole in the ozone layer is heating up the planet.') (Leiserowitz, A., Smith, N., and Marlon, J.R. 2010). Unreliable sources in the media and popular culture generate, magnify, and proliferate misconceptions. For example, studies (Cook et al. 2016) show that there is overwhelming consensus among climate scientists that climate change is happening and is primarily due to human activity. But among the American public, just over half believed this to be the case in 2017 (Ballew et al. 2019); in 2021, this number is up to 57% according to Yale Climate Opinion Maps.[1] The funding of skeptic groups by fossil fuel companies (Cook, J. et al. 2019) and the attempt by big corporations to influence and derail climate awareness and action (Oreskes 2011) have further obscured the issue in the public mind. According to a recent research report (Nadeem 2020), despite bipartisan support for alternative energy and environmental protection, 'Americans continue to be deeply politically divided about how much human activity contributes to climate change.' The politicization of climate change, leading to widespread climate denialism, has become a barrier to effective climate education and action in several conservative US states because of its influence on school curricula, textbooks, and policies. A groundbreaking study in 2016 (Plutzer et al. 2016) titled 'Climate Confusion Among US Teachers' points to a significant proportion of US teachers presenting mixed messages about the cause of climate change (denying or underplaying the role of human activity) while other work (Berbeco, Heffernan, and Branch 2017) explores the impact of such denial and doubt in the classroom. More recent studies reveal the impact of misleading and denialist material in textbooks on adolescents' certainty about climate change (Busch 2021), and the role of oil corporations, school boards, state legislatures, and conservative think tanks in influencing what is taught and how it is taught

(Worth, Katie 2021). An online survey[2] by YouGov in 2020 of 26,000 people in 25 countries indicated that the countries with the highest proportion of climate deniers are the United States and Indonesia, at 21% and 19%, respectively, closely followed by Saudi Arabia and Egypt at 18% and India at 16%. An online survey is necessarily nonrepresentative, especially in the Global South, where most people do not have access to the Internet; however, within its limitations, the survey indicates that climate denialism is not limited to one country or region.

Some confusion also arises because education has failed to communicate how science works, and in particular, the meaning and role of uncertainty. Science by its very nature is provisional and subject to constant revision and correction. Climate science has specific challenges that result in some uncertainties that are inherent and others that represent a lack of knowledge (Risbey and O'Kane 2011). However, the basic science of climate change, resting on thermodynamics, the physics of greenhouse gases, and the carbon cycle, supports unequivocally the reality of anthropogenic climate change (IPCC 2023). But reports of uncertainties in climate projections, when not adequately explained and contextualized, can lead the public to conclude that the reality of anthropogenic climate change is in question. These considerations lead us to ask the question: what is essential climate change knowledge? Here, by 'essential,' I mean knowledge that has explanatory power with regard to the myriad biophysical changes taking place, is robust against disinformation, and is a prerequisite to wise action. This is elucidated in detail in Chapter 4.

2   The challenge of inter-/transdisciplinarity: As the stories in the introductory chapter indicate, climate change is a phenomenon at the intersection of physics, chemistry, biology, economics, sociology, psychology, and Indigenous rights, to name but a few areas – the ultimate 'wicked problem.' Our siloed system of education does not easily allow space for a truly inter/transdisciplinary exploration of climate change. Disciplines have developed specific lexicons, paradigms, and frameworks that may not be easily translatable across boundaries; hence, transdisciplinary scholarship is a relatively new field (Brown, V.A., Harris, J., and Russell, J. 2010; Leavy 2011). This challenge is both cognitive and structural. Thinking across disciplines requires an epistemological shift (item 4 below) on the part of both instructor and student. My experience co-conceptualizing and co-running a week-long interdisciplinary workshop for middle and high school science teachers in Massachusetts in 2017 indicated that barriers between disciplines represented the greatest cognitive challenge, despite the participants' openness to the idea. Scholars point out that differences in values as well as epistemologies can confound interdisciplinary collaboration and thinking (Lélé and Norgaard 2005). The lack of proper teacher training is a worldwide problem and is discussed in item 5 below.

3   Psychological barriers – the psychology of climate change is still in its infancy but it has been suggested (Wysham, Daphne 2012) that learning about climate change results in emotional trauma that includes denial, anger, and despair. This is consistent with my teaching experience and that of multiple researchers who point out the crucial importance of attending to students' affective responses to learning about the climate crisis (Hufnagel 2017; Verlie 2019). Climate change and its implications are frightening. The dilemma is that while we, as educators, are obligated to tell the truth, the truth can be an overwhelming emotional burden. Science educators are not equipped to handle the affective impact of their teaching upon students. Being aware that climate change is an emotionally fraught issue (APA Taskforce on Interface Between Psychology and Global Climate Change 2009), we can apply ways to work through this aspect that allows the student to engage meaningfully with the subject.

4   Onto-Epistemological Barriers – As expressed in Table 2.1, Modern industrial cultures' constructs of reality, what we might call the dominant paradigm, does not fit the characteristic features of the climate crisis. Thus 'solutions' arising from the same paradigm or worldview that brought us the climate crisis are unlikely to be truly effective in the long term. As I indicate in Table 2.1 and elaborate in Chapter 4, a key aspect of the climate problem is complex interconnections across space, time, societies, and species, a complexity that is either oversimplified or ignored in the dominant paradigm. The mechanistic or Newtonian way of seeing the world, a key aspect of modern industrial cultures, washes out these essential connections so that we exist in fragmented spaces, times, and relationships.

Table 2.1 Key features of the climate crisis contrasted with broad-brush aspects of modern industrial societies

Key Features of the Climate Crisis	Broad-Brush Aspects of Modern Industrial Societies
Spans large scales of space and time	Limited spatial and temporal scales and fragmentation of space and time
Is inherently transdisciplinary and connected with other major social-environmental problems	Fragmentation of knowledge into disciplines that exist in silos with little interaction; problems are seen as reductive and separate
Is rife with complex interconnections at multiple scales, across space and time, and between and among human societies and the rest of Nature	Simple linear causality dominates thinking; world seen primarily as mechanistic, simple; little or no recognition that we live within multiple interconnected social-natural complex systems; Nature-Culture divide and anthropocentrism are defaults
Is centrally a problem of justice and power	Living in hierarchical societies with major power imbalances that encourage a blindness to inequality and considerations of justice

These present a barrier of the imagination that can often prevent alternative ways to conceptualize, think, and act. This onto-epistemological blindness is perhaps the greatest cognitive barrier to understanding and acting effectively on the climate crisis. Of course, the term 'dominant paradigm' conceals nuances of culture, history, and geography, so it is important to recognize it as a coarse-grained, broad-brush generalization. The term 'dominant' has a special meaning for settler-colonized regions of the world and formerly colonized nations because the imposition of colonial, typically Eurocentric ideas, concepts, and sense-making has often suppressed or sidelined alternative cultural conceptualizations and responses to the climate problem (Reibold n.d.; Gwekwerere and Shumba 2021; Whyte 2021; Bhambra and Newell 2022; Bogert et al. 2022). The wholesale adoption of the term 'Anthropocene' is but one example. This dominance manifests in the education system of countries that were once colonized, such as in India (Shin and Akula 2021; Raveendran and Srivastava 2022). The dominance of Eurocentric and particularly neoliberal formulations of environmental problems can lead to uncritical acceptance of techno-managerial, top-down approaches implied by concepts such as Education for Sustainable Development, and the UN's Sustainable Development Goals (Bogert et al. 2022). It's not that these don't have some useful and important ideas but the hegemonic exclusion of alternatives is, as I hope to show, highly problematic. The US NGSS for example has been critiqued (Clark, Sandoval, and Kawasaki 2020) as favoring universalism (presenting climate change as global without connecting with and foregrounding its local manifestations) as well as scientism and technocentrism. An unfolding epistemological expansion in climate science appears to recognize the problematic aspects of a top-down, purely technocentric approach to the work of climate scientists with local communities (Shepherd and Lloyd 2021). It is important for educators to be aware of the changing epistemological landscape of climate science.

Thus, the project of restoring ontological and epistemological justice in knowledge-making inevitably implies challenging standard teaching practices, teacher-student relationships, academic hierarchies, and the place of academia in society. Even within the classroom, an onto-epistemological re-orientation requires an epistemic shift on the part of teachers before it can be enabled for the student. Multiple scholars of transformative education have pointed out that an epistemic shift must undergird any effective pedagogy of social-environmental problems (Mezirow, J. and Taylor, E.W. 2009; Sterling 2011; Lotz-Sisitka et al. 2015; Boström et al. 2018; Macintyre et al. 2018; Odell et al. 2020). However, a shortcoming of most approaches is that the focus is on an epistemic shift in an individual student, ignoring the impact of relationships with other students and the educator, and the role of the classroom as community. In general, conventional institutions of learning are difficult places to engender transformative learning (Sterling 2011).

5 Teacher Training and Institutional Barriers

The above challenges make it imperative for teachers in all disciplines to receive training in transdisciplinary and transformative education methodologies. This would entail collaboration between far-apart disciplines such as physics and sociology, or biology and literature. The rigid structural barriers in educational institutions effectively wall off disciplines from each other; there is generally little recognition or appreciation for the value of cross-disciplinary training and collaboration. Thus, at this point, educators wishing to teach climate change across disciplines have to teach themselves. Even within disciplines, teachers report the difficulties of rigid, top-down administrative structures that allow little room for innovation; limitations of prescribed national curricula, undue emphasis on examinations that encourage 'teaching to the test,' and a general lack of understanding of the importance of environmental education throughout the educational hierarchy that results in sustainability and climate education being relegated to the margins or considered add-ons or restricted to science classes. Additionally, lack of adequate or high-quality teacher training makes teachers feel underconfident about teaching these topics. Various studies indicate that this is the case in countries of the Global South as well as North America and Europe (Monroe et al. 2019; González 2021; Gwekwerere and Shumba 2021). In the Global South, there are also problems of teacher shortage and lack of resources for effective teaching. An examination of the Indian context (Shin and Akula 2021; Raveendran and Srivastava 2022) reveals the onto-epistemological dominance of Eurocentric conceptualizations of education in general and environment in particular.

In my own case, my training is in particle physics, so I had to take the initiative to learn basic climate science (an ongoing journey), and to ask myself: from the vast and rapidly developing field of climate science: what concepts can be considered essential to a lay but sophisticated understanding of the crisis? Since I could only teach climate change in the context of a physics class, I had to figure out these essentials and learn how to connect them with existing course topics. I also had to learn from sociologists, economists, anthropologists, historians, and scholars of environmental humanities, among others, and integrate key ideas from those disciplines into my framework. From the logistical challenge of teaching climate change in a classroom devoted to another subject (even with a transdisciplinary approach), there emerged a significant problem: the aforementioned *piecemeal learning*, where climate topics appear at irregular intervals during the course in disconnected chunks. Such piecemeal learning would prevent an integrated, comprehensive 'big picture' understanding of the material, and undermine the intended purpose.

If we are to take seriously the teachings of the climate problem, then any effective pedagogy must address these barriers, and in doing so, embrace the key features of the problem: its transdisciplinarity, its spanning of large scales of space and time, its rich complexity, and its roots in injustice and power. But how to realize this, when it goes against the grain of much of mainstream education? How to do better than a few radical visionary educators working in their microspaces, whose impact will necessarily be very limited?

This last is too large a question to address in this book. Suffice it to say that what is really needed is large-scale educational restructuring, in which climate change and other problems are taught in a way that does justice to them, and allows students and educators to be among the changemakers for a viable future. This radical restructuring would address power imbalances and bring down the walls between the classroom and the community-at-large as well as between disciplines. As we (hopefully) work toward the kind of radical change warranted by the seriousness of our social-environmental problems, we must resort, in the short term, to teaching climate change effectively, both individually and across the curriculum. This requires lasting institutional support of faculty and students.

Granting that change is required across scales, let me narrow down the question to the context from which my explorations have risen: a single classroom focused on a subject that is not climate science (in my case, general physics). No matter the subject, is it possible to teach climate change in a way that responds robustly to its key teachings? What would such an approach involve?

2.4   What Is an Effective Pedagogy of Climate Change?

We must then ask: what do we want students to take away from their study of climate change? What is an effective pedagogy of climate change? This is not a simple question. The common answer from the standard perspective of a science educator is that we need students to understand the science and the implications, and that is where our responsibility ends. Consistent with my experience, studies show that knowing how serious the implications are does not result in action, rather it can impede action (Chess and Johnson 2007; Kahan et al. 2012). The pervasive argument that scientists should simply do the science (and by implication, teach just the science) is no longer tenable in the era of climate change and related crises.

I propose that an effective pedagogy of climate change in a physics (or other) classroom is one that embraces the key characteristics (teachings) of the problem and empowers the learners to take part in meaningful action. That is, such an approach:

a   Equips the student with a fundamental understanding of the basic science, impacts, and evidence of climate change, including its complex, nonlinear nature, as well as the future projections based on various scenarios – *the scientific-technological dimension* b   Enables the student to understand societal and ethical implications of climate change, leading to intersections with economic, cultural, human rights, and sociological issues; to understand how climate change is related to other major social-ecological problems and to critically examine proposed climate solutions from a climate justice perspective – *the transdisciplinary dimension*

c   Enables the student to see the climate crisis as a symptom of a social-scientific framework or paradigm; to recognize the key features of the dominant paradigm of modern industrial society and its historical relationship to colonialism, so as to understand and articulate the need for new alternative social-scientific frameworks in order to usefully engage with the crisis – *the epistemological dimension*

d   Through their own epistemic shift, inspires students to explore their own affective and cognitive responses to the crisis, as well as their agency, and motivates them to engage collectively with social-environmental problems in society, and to recognize and confront inequality, injustice and power hierarchies – *the psychosocial-action dimension*

Such a pedagogy, when fully realized and implemented, is consistent with Science Literacy Vision III (Sjöström and Eilks 2018) and Critical Scientific Literacy (Hodson 2010) in that it goes beyond an examination of the social context of science to socio-political action, recognizing and challenging power hierarchies.

Although the above are at the classroom level rather than program level, they further the development of four of five key competencies identified for sustainability education: systems-thinking, anticipatory, normative and interpersonal competencies (Wiek, A., Withycombe, L., and Redman, C. 2011). These four dimensions are not separate, but overlap, and justice considerations constitute the connective tissue of this approach, as I will demonstrate in subsequent chapters.

An effective pedagogy must necessarily develop and apply best practices for cognitive and affective understanding. This requires a change in classroom culture from the conventional lecture-heavy, teacher as sage-on-stage approach. My teaching is inspired by the work of Carol Dweck on mindsets (Dweck, C.S. 2006), and of Ken Bain on the Natural Critical Learning Environment (Bain 2004; Bain, K. and Zimmerman, J. 2009) as well as Embodied Learning (Euler, Rådahl, and Gregorcic 2019). More recently, I have discovered that my approach is consistent with the ethos of transformational learning (for an overview, see Hoggan, C.D. 2018). A full realization of all four dimensions should ideally enable students to undergo an epistemic shift (Mezirow, J. and Taylor, E.W. 2009). Sterling, following Bateson, makes a very interesting distinction between three orders of change and learning: conformative change, focused on effectiveness and efficiency, reformative change, focused on examining and changing assumptions and transformative change that results in an epistemic shift (Sterling 2011): 'learning that facilitates a fundamental recognition of paradigm and enables paradigmatic reconstruction.'

I have been working toward such an approach and improving on it for the past decade, in all levels of undergraduate general physics courses, in particular Physics, Nature, and Society, a physics course with laboratory for nonscience majors. This pedagogy has also been employed in a First Year Seminar on Arctic climate change. I start with posing an overarching question or theme for the course before the course begins, which is sent out as an invitation to students (Bain 2004).

For most of my general physics courses, the theme has consistently been climate change. The courses are taught with an active learning focus – interactive lectures are interspersed with think-pair-share exercises, group work, student exploration with laboratory equipment, embodied learning in the form of 'physics theater,' whereby students work in groups to enact physical principles, chances for students to hypothesize without worrying about being wrong, second chances on certain exams, and a focus on each student – that is, helping *every* student toward excellence through extra help and encouragement. A consistent effort is made to build trust between instructor and students, and to create an environment where students feel psychologically safe and valued so that they can be intellectually audacious. I have found the use of microaffirmations (Rastegari, Iman and Shafer, Leah 2016) to be especially powerful for a diverse, mixed-race classroom. One of the implications is that even if the student appears to be unmotivated, does assignments poorly, doesn't attend class regularly, and is generally a 'poor student,' I *still* interact with that student in a positive, welcoming, and caring manner. It is not uncommon for educators to regard such students with dismissive scorn, but such an attitude is judgmental, misinformed, and counter-productive. In addition, I make a deliberate attempt to flatten the power hierarchy in the classroom – for instance, every week to two weeks, students have a chance to give me anonymous feedback on their experience of the course. I ask them to report what's working for them, what I could do better, and what questions and concerns they might have. At the start of next class, we have a discussion on the feedback. This helps me keep a finger on the pulse of the classroom, adapt my teaching approach to better serve students, clarify my perspective and appreciate theirs, enable community decision-making, and most importantly, *help students realize that they matter.* Early in my teaching career, a student's comment that they found my blackboard work confusing helped me to become more systematic in presenting concepts and mathematical equations on the blackboard. In my end-semester surveys, students often report that they feel heard in my classroom – for example, one student wrote 'you really listen to us!' I know I am successful when students begin to speak up in class with their views and concerns without waiting for the anonymous feedback opportunity. I see this also when previously shy or apparently unmotivated students begin to ask questions in class, seek extra help, and point out inadvertent mistakes in algebra that I might have made on the blackboard!

When I have discussed my approach with colleagues, some of them have immediately jumped to the conclusion this welcoming approach that seeks to empower student voices in the classroom necessarily means lowering standards. After all, science requires rigor, and if you either let students 'do whatever they want' or go on dubious interdisciplinary excursions, you must surely be diluting the science. However, nothing could be further from the truth. Because this reaction is so widespread, it warrants a detailed refutation.

Typically, science classrooms – in physics especially – are considered 'weed-out' spaces, where those who cannot keep up with the pace and rigor are discarded along the way, and only the chosen few remain. This presupposes that there are only a few people with the brainpower to really understand the material. (The patriarchal male scientist-genius mystique in popular culture doesn't help.) Thus, I have come across mathematics and physics professors who have no hesitation in telling a student that 'you are not capable of understanding this.'

However, work by Carol Dweck and others (Yeager and Walton 2011; Dweck 2015) indicates that student success is based on multiple factors, of which the psychosocial seem to be significant – for example, stereotype threat among women and minoritized students (Totonchi et al. 2021) and the importance of the feeling of belonging (Broda et al. 2018) among under-represented groups and first-generation college goers. Dweck's work identifies mindset as a key factor in student success – a fixed mindset assumes that traits such as intelligence are predetermined and fixed at birth (hence the custom in the United States of marking some students as 'gifted and talented') while a growth mindset considers these traits to be a function of hard work and the proper application and development of one's mind to the task, including an attitude of curiosity and openness. Many cultures seem to enforce a fixed mindset toward learning and intelligence on their young, through labels such as 'below average,' 'average,' 'smart,' etc., all of which are limiting and potentially harmful. But mindsets can be changed, as Dweck's work shows, with powerful consequences. In my own experience, I have had the privilege of helping numerous students realize their intellectual potential – engaging their curiosity, building their confidence through multiple one-on-one sessions, and having them see their grades go from D and F to A and B. This transformation is only possible *when we don't dumb down the course material to the lowest common denominator but instead insist on high standards*. To demonstrate faith in a student's ability, we *must* have high standards and high expectations, and then do the hard work of helping students reach those standards. I tell my students that in the natural sciences, the ultimate authority is not the scientist in the room, or, in fact, any human being, but Nature. Thus I, myself, despite my authority as professor, am a learner in Nature's classroom, which is the universe. This also allows us to teach students to learn from failure, because, in a sense, failure is a way for Nature to tell us there are surprises, that we need to think harder, dig deeper. An answer to a question in the classroom such as 'what happens if we drop this heavy ball and this light ball from the same height at the same time' that is contrary to the result of the experiment is simply Nature telling us to think again. Therefore, failures can invoke curiosity and invite further exploration, if we look at these moments as teachers. I emphasize this point throughout the semester, pointing out how wrong turns and incorrect hypotheses in my own Ph.D. research helped me find the answer to my research question. I help them analyze their returned, graded work, so that they know how to do better next time, and give them multiple chances to succeed. I give them training in best practices for learning (Brown, Peter C., Roediger, Henry L. III, and McDaniel, Mark A. 2014; McGuire, Saundra Yancy 2018) to develop their metacognitive skills and help them through despair, frustration, stumbles, and difficult times. This involves a great deal of hard work (generally unacknowledged by administration) but it is immensely rewarding. To cite just one example, in one upper level physics class for science majors, a student who initially struggled with problem-solving was ultimately able to successfully solve on his own, problems I had borrowed from a similar MIT course. For an African-American first-generation student who had doubted his ability to pass the course, this was a powerful turning point. And it would not have happened without my insistence on high standards backed by the relationship building between myself and the students, as well as the exercises in empowerment and community building. Therefore, to the doubters, my response is *Rigor, but not Rigor Mortis*. My classroom culture is thus diametrically opposed to the 'weed-out' culture and is, in fact, dedicated to the success of every student, without exception. While this does not always work in practice, with some students slipping through the cracks despite my efforts, I always let them know that even if they fail the course, I will not lose faith in them. My pilot studies in these classes indicate improvement in confidence, interest, and performance among a majority of students (80–96% based on exit surveys).

Finally, since I have mentioned physics theater and enactment (elaborated in later chapters), and because learning is an embodied process, I must add a note about the physical space of the classroom. The traditional rows of chairs and desks, all facing the teacher's podium, encourage the notion that students are passive recipients of pre-digested knowledge, with the teacher as the 'sage on the stage.' This is completely antithetical to the kind of active, critical, collaborative classroom culture that I hope to build. Therefore, for many of my classes, I have been using a room where the furniture arrangement consists of five large, pentagonal tables, each able to seat five, each with mobile, wheeled chairs. The tables are set along the classroom walls, and there is a clear square space in the middle that serves as a collaborative space. The tables allow me to place materials for students to play with, such as pendulums to investigate resonance, and the central space is a stage and a gathering place. Coming into such a classroom upends the unconscious expectation of passive learning.

Since teaching climate change is beset with challenges additional to those that we encounter in the general physics classroom, it is important to employ these best practices for teaching. In subsequent chapters, I will lay out how the four dimensions of an effective climate pedagogy might be realized in such a classroom.

This leads to an important observation and a possible drawback of my approach, and indeed any new approach that departs sufficiently from conventional teaching: many standard measures of efficacy may be irrelevant or inapplicable. We do not, by definition, have 'standard measures' for non-standard approaches. Take transdisciplinarity, which is one way that my approach departs from convention – there are so many ways to be transdisciplinary (and the falling of disciplinary walls is still such a new phenomenon where it exists) that to imagine one standard, quantitative assessment for all these is absurd. There are qualitative approaches that I have learned about through the generosity of colleagues in the social sciences, such as coding written responses from students, and I have, on occasion, employed these techniques. They have their place; however, in pedagogical explorations such as mine, they seem to leave out more than they capture. In later chapters, I introduce a different paradigm, which has less to do with formal assessment and more to do with generating new insights by considering relationships and 'intra-actions' within the classroom and beyond, within disciplines and outside them. This paradigm of *diffractive analysis*, originating from the work of physicist and feminist philosopher Karen Barad (2007), is elaborated in the final chapter of this book.

Notes

1 https://climatecommunication.yale.edu/visualizations-data/ycom-us/
2 https://www.statista.com/chart/19449/countries-with-biggest-share-of-climate-change-deniers/

References

APA Taskforce on Interface Between Psychology and Global Climate Change. 2009. *Psychology and Global Climate Change: Addressing a Multifaceted Phenomenon and Set of Challenges*. Washington, DC: American Psychological Association.

Bain, K., and J. Zimmerman 2009. "Understanding Great Teaching." Text. Association of American Colleges & Universities. https://www.aacu.org/publications-research/periodicals/understanding-great-teaching.

Bain, Ken. 2004. "What Makes Great Teachers Great?" *The Chronicle of Higher Education: The Chronicle Review* 50 (31): B7.

Ballew, Matthew T., Anthony Leiserowitz, Connie Roser-Renouf, Seth A. Rosenthal, John E. Kotcher, Jennifer R. Marlon, Erik Lyon, Matthew H. Goldberg, and Edward W. Maibach. 2019. "Climate Change in the American Mind: Data, Tools, and Trends." *Environment: Science and Policy for Sustainable Development* 61 (3): 4–18. https://doi.org/10.1080/00139157.2019.1589300.

Barad, Karen. 2007. *Meeting the Universe Halfway: Quantum Physics and the Entanglement of Matter and Meaning*. Durham, NC: Duke University Press.

Berbeco, Minda, Kate Heffernan, and Glenn Branch. 2017. "Doubt and Denial as Challenges to, and in, Teaching Climate Change." In *Teaching and Learning about Climate Change*. Abingdon-on-Thames, UK: Routledge.

Bhambra, Gurminder K., and Peter Newell. 2022. "More than a Metaphor: 'Climate Colonialism' in Perspective." *Global Social Challenges Journal* 1 (aop): 1–9. https://doi.org/10.1332/EIEM6688.

Bogert, Jeanne, Jacintha Ellers, Stephan Lewandowsky, Meena Balgopal, and Jeffrey Harvey. 2022. "Reviewing the Relationship between Neoliberal Societies and Nature: Implications of the Industrialized Dominant Social Paradigm for a Sustainable Future." *Ecology and Society* 27 (2). https://doi.org/10.5751/ES-13134-270207.

Boström, Magnus, Erik Andersson, Monika Berg, Karin Gustafsson, Eva Gustavsson, Erik Hysing, and Rolf Lidskog, et al. 2018. "Conditions for Transformative Learning for Sustainable Development: A Theoretical Review and Approach." *Sustainability* 10 (12): 4479. https://doi.org/10.3390/su10124479.

Broda, Michael, John Yun, Barbara Schneider, David S. Yeager, Gregory M. Walton, and Matthew Diemer. 2018. "Reducing Inequality in Academic Success for Incoming College Students: A Randomized Trial of Growth Mindset and Belonging Interventions." *Journal of Research on Educational Effectiveness* 11 (3): 317–38. https://doi.org/10.1080/19345747.2018.1429037.

Brown, Peter C., Henry L. Roediger III, and Mark A. McDaniel. 2014. *Make It Stick*. Cambridge, MA: Belknap Press.

Brown, V. A., J. Harris, and J. Russell. 2010. *Tackling Wicked Problems Through the Transdisciplinary Imagination*. Abingdon-on-Thames: Routledge.

Busch, K.C. 2021. "Textbooks of Doubt, Tested: The Effect of a Denialist Framing on Adolescents' Certainty about Climate Change." *Environmental Education Research* 27 (11): 1574–98. https://doi.org/10.1080/13504622.2021.1960954.

Chess, Caron, and Branden B. Johnson. 2007. "Information Is Not Enough." In *Creating a Climate for Change: Communicating Climate Change and Facilitating Social Change*, edited by Lisa Dilling and Susanne C. Moser, 223–34. Cambridge: Cambridge University Press. https://doi.org/10.1017/CBO9780511535871.017.

Clark, Heather F., William A. Sandoval, and Jarod N. Kawasaki. 2020. "Teachers' Uptake of Problematic Assumptions of Climate Change in the NGSS." *Environmental Education Research* 26 (8): 1177–92. https://doi.org/10.1080/13504622.2020.1748175.

Cook, J., G. Supran, S. Lewandowsky, N. Oreskes, and E. Maibach. 2019. *America Misled: How the Fossil Fuel Industry Deliberately Misled Americans about Climate Change*. Fairfax, VA: George Mason University Center for Climate Change Communication.

Cook, John, Naomi Oreskes, Peter T Doran, William R L Anderegg, Bart Verheggen, Ed W Maibach, and J Stuart Carlton, et al. 2016. "Consensus on Consensus: A Synthesis of Consensus Estimates on Human-Caused Global Warming." *Environmental Research Letters* 11 (4): 048002. https://doi.org/10.1088/1748-9326/11/4/048002.

Dweck, Carol. 2015. "Carol Dweck Revisits the 'Growth Mindset.'" *Education Week*, September 23, 2015, sec. Leadership, Student Well-Being. https://www.edweek.org/leadership/opinion-carol-dweck-revisits-the-growth-mindset/2015/09.

Dweck, C.S. 2006. *Mindset: The New Psychology of Success*. New York, NY: Random House.

Euler, Elias, Elmer Rådahl, and Bor Gregorcic. 2019. "Embodiment in Physics Learning: A Social-Semiotic Look." *Physical Review Physics Education Research* 15 (1): 010134. https://doi.org/10.1103/PhysRevPhysEducRes.15.010134.

González, Estefanía Pihen. 2021. *Toward Education for Sustainable Development: Lessons from Asia and the Americas.* Brill. https://doi.org/10.1163/9789004471818_019.

Gwekwerere, Yovita N., and Overson Shumba. 2021. *A Call for Transformative Learning in Southern Africa: Using Ubuntu Pedagogy to Inspire Sustainability Thinking and Climate Action.* Brill. https://doi.org/10.1163/9789004471818_011.

Hodson, Derek. 2010. "Science Education as a Call to Action." *Canadian Journal of Science, Mathematics and Technology Education* 10 (3): 197–206. https://doi.org/10.1080/14926156.2010.504478.

Hoggan, C.D. 2018. "The Current State of Transformative Learning Theory: A Metatheory." *Phronesis* 7 (3): 18–25. https://doi.org/10.7202/1054405ar.

Hufnagel, Elizabeth. 2017. "Attending to Emotional Expressions about Climate Change: A Framework for Teaching and Learning." In *Teaching and Learning about Climate Change.* Abingdon-on-Thames, UK: Routledge.

Huntington, Henry P., Andrey Zagorsky, Bjørn P. Kaltenborn, Hyoung Chul Shin, Jackie Dawson, Maija Lukin, Parnuna Egede Dahl, Peiqing Guo, and David N. Thomas. 2022. "Societal Implications of a Changing Arctic Ocean." *Ambio* 51 (2): 298–306. https://doi.org/10.1007/s13280-021-01601-2.

IPCC. 2023. "Synthesis Report of the IPCC Sixth Assessment Report (AR6)." https://www.ipcc.ch/report/sixth-assessment-report-cycle/.

Kahan, Dan M., Ellen Peters, Maggie Wittlin, Paul Slovic, Lisa Larrimore Ouellette, Donald Braman, and Gregory Mandel. 2012. "The Polarizing Impact of Science Literacy and Numeracy on Perceived Climate Change Risks." *Nature Climate Change* 2 (10): 732–35. https://doi.org/10.1038/nclimate1547.

Kwauk, C. 2020. "Roadmaps to Quality Education in a Time of Climate Change." Brief. Washington, DC: Brookings Institute. https://www.brookings.edu/wp-content/uploads/2020/02/Roadblocks-to-quality-education-in-a-time-of-climate-change-FINAL.pdf.

Leavy. 2011. *Essentials of Transdisciplinary Research.* 1st ed. Abingdon-on-Thames Routledge.

Lee, Tien Ming, Ezra M. Markowitz, Peter D. Howe, Chia-Ying Ko, and Anthony A. Leiserowitz. 2015a. "Predictors of Public Climate Change Awareness and Risk Perception around the World." *Nature Climate Change* 5 (11): 1014–20. https://doi.org/10.1038/nclimate2728.

———. 2015b. "Predictors of Public Climate Change Awareness and Risk Perception around the World." *Nature Climate Change* 5 (11): 1014–20. https://doi.org/10.1038/nclimate2728.

Leiserowitz, A., J. Carman, N. Buttermore, L. Neyens, S. Rosenthal, J. Marlon, J Schneider, and K. Mulcahy 2022. "International Public Opinion on Climate Change, 2022." New Haven, CT: Yale Program on Climate Change Communication, Data For Good at Meta. https://climatecommunication.yale.edu/publications/international-public-opinion-on-climate-change-2022/.

Leiserowitz, A., N. Smith, and J.R. Marlon 2010. "Americans' Knowledge of Climate Change." Yale Project on Climate Change Communication. New Haven, CT: Yale University. https://environment.yale.edu/climate-communication-OFF/files/ClimateChangeKnowledge2010.pdf.

Lélé, Sharachchandra, and Richard B. Norgaard. 2005. "Practicing Interdisciplinarity." *BioScience* 55 (11): 967–75. https://doi.org/10.1641/0006-3568(2005)055[0967:PI]2.0.CO;2.

Lipson, Sarah Ketchen, Sasha Zhou, Sara Abelson, Justin Heinze, Matthew Jirsa, Jasmine Morigney, Akilah Patterson, Meghna Singh, and Daniel Eisenberg. 2022. "Trends in College Student Mental Health and Help-Seeking by Race/Ethnicity: Findings from the National Healthy Minds Study, 2013–2021." *Journal of Affective Disorders* 306 (June): 138–47. https://doi.org/10.1016/j.jad.2022.03.038.

Lotz-Sisitka, Heila, Arjen EJ Wals, David Kronlid, and Dylan McGarry. 2015. "Transformative, Transgressive Social Learning: Rethinking Higher Education Pedagogy in Times of Systemic Global Dysfunction." *Current Opinion in Environmental Sustainability* 16: 73–80. https://doi.org/10.1016/j.cosust.2015.07.018.

Macintyre, Thomas, Heila Lotz-Sisitka, Arjen Wals, Coleen Vogel, and Valentina Tassone. 2018. "Towards Transformative Social Learning on the Path to 1.5 Degrees." *Current Opinion in Environmental Sustainability* 31 (April): 80–87. https://doi.org/10.1016/j.cosust.2017.12.003.

McGuire, Saundra Yancy. 2018. *Teach Yourself How to Learn.* Sterling, VA: Stylus Publishing. https://styluspub.presswarehouse.com/browse/book/9781620367568/Teach-Yourself-How-to-Learn.

Mezirow, J., and E.W. Taylor, eds. 2009. *Transformative Learning in Practice: Insights from Community, Workplace, and Higher Education.* Hoboken, NJ: Wiley.

Monroe, Martha C., Richard R. Plate, Annie Oxarart, Alison Bowers, and Willandia A. Chaves. 2019. "Identifying Effective Climate Change Education Strategies: A Systematic Review of the Research." *Environmental Education Research* 25 (6): 791–812. https://doi.org/10.1080/13504622.2017.1360842.

Nadeem, Reem. 2020. "Two-Thirds of Americans Think Government Should Do More on Climate." *Pew Research Center Science & Society* (blog). June 23, 2020. https://www.pewresearch.org/science/2020/06/23/two-thirds-of-americans-think-government-should-do-more-on-climate/.

Odell, Vanessa, Petra Molthan-Hill, Stephen Martin, and Stephen Sterling. 2020. "Transformative Education to Address All Sustainable Development Goals." In *Quality Education*, edited by Walter Leal Filho, Anabela Marisa Azul, Luciana Brandli, Pinar Gökçin Özuyar, and Tony Wall, Encyclopedia of the UN Sustainable Development Goals, 905–16. Cham: Springer International Publishing. https://doi.org/10.1007/978-3-319-95870-5_106.

Oreskes, Naomi. 2011. *Merchants of Doubt : How a Handful of Scientists Obscured the Truth on Issues from Tobacco Smoke to Global Warming.* 1st U.S. edition. New York, NY: Bloomsbury Press. ©2010. https://search.library.wisc.edu/catalog/9910089279802121.

Plutzer, Eric, Mark McCaffrey, A. Lee Hannah, Joshua Rosenau, Minda Berbeco, and Ann H. Reid. 2016. "Climate Confusion among U.S. Teachers." *Science* 351 (6274). 664–65. https://doi.org/10.1126/science.aab3907.

Rastegari, Iman, and Leah Shafer. 2016. "Accentuate the Positive." Harvard Graduate School of Education. December 22, 2016. https://www.gse.harvard.edu/news/uk/16/12/accentuate-positive.

Raveendran, Aswathy, and Himanshu Srivastava. 2022. "Science and Environment Education in the Times of the Anthropocene: Some Reflections from India." In *Reimagining Science Education in the Anthropocene*, edited by Maria F. G. Wallace, Jesse Bazzul, Marc Higgins, and Sara Tolbert, 201–13. Palgrave Studies in Education and the Environment. Cham: Springer International Publishing. https://doi.org/10.1007/978-3-030-79622-8_12.

Reibold, Kerstin. n.d. "Settler Colonialism, Decolonization, and Climate Change." *Journal of Applied Philosophy* n/a (n/a). Accessed March 31, 2023. https://doi.org/10.1111/japp.12573.

Risbey, James S., and Terence J. O'Kane. 2011. "Sources of Knowledge and Ignorance in Climate Research." *Climatic Change* 108 (4): 755. https://doi.org/10.1007/s10584-011-0186-6.

Shepherd, Theodore G., and Elisabeth A. Lloyd. 2021. "Meaningful Climate Science." *Climatic Change* 169 (1): 17. https://doi.org/10.1007/s10584-021-03246-2.

Shin, Haein, and Srinivas Akula. 2021. *Educators' Perspectives on Environmental Education in India: A Case Study in School and Informal Education Settings.* Brill. https://doi.org/10.1163/9789004471818_018.

Sjöström, Jesper, and Ingo Eilks. 2018. "Reconsidering Different Visions of Scientific Literacy and Science Education Based on the Concept of Bildung." In *Cognition, Metacognition, and Culture in STEM Education: Learning, Teaching and Assessment*, edited by Yehudit Judy Dori, Zemira R. Mevarech, and Dale R. Baker, 65–88. Innovations in Science Education and Technology. Cham: Springer International Publishing. https://doi.org/10.1007/978-3-319-66659-4_4.

Sterling, Stephen. 2011. "Transformative Learning and Sustainability: Sketching the Conceptual Ground." *Learning and Teaching in Higher Education* 5: 17–33.

Thompson, Tosin. 2021. "Young People's Climate Anxiety Revealed in Landmark Survey." *Nature News*, September 22, 2021. https://www.nature.com/articles/d41586-021-02582-8.

Totonchi, Delaram A., Tony Perez, You-kyung Lee, Kristy A. Robinson, and Lisa Linnenbrink-Garcia. 2021. "The Role of Stereotype Threat in Ethnically Minoritized Students' Science Motivation: A Four-Year Longitudinal Study of Achievement and Persistence in STEM." *Contemporary Educational Psychology* 67 (October): 102015. https://doi.org/10.1016/j.cedpsych.2021.102015.

Verlie, Blanche. 2019. "'Climatic-Affective Atmospheres': A Conceptual Tool for Affective Scholarship in a Changing Climate." *Emotion, Space and Society* 33 (November): 100623. https://doi.org/10.1016/j.emospa.2019.100623.

Verlie, Blanche. 2021. *Learning to Live With Climate Change: From Anxiety to Transformation.* Abingdon, Oxon: Routledge.

Whyte, Kyle Powys. 2021. "Time as Kinship." In *The Cambridge Companion to Environmental Humanities*, edited by Jeffrey Cohen and Stephanie Foote, 39–55. Cambridge Companions to Literature. Cambridge: Cambridge University Press. https://doi.org/10.1017/9781009039369.005.

Wiek, A., L. Withycombe, and C. Redman 2011. "Key Competencies in Sustainability: A Reference Framework for Academic Program Development." *Sustainability Science* 6: 203–18. https://doi.org/10.1007/s11625-011-0132-6.

Worth, Katie. 2021. *Miseducation: How Climate Change Is Taught in America.* New York, NY: Columbia Global Reports. https://globalreports.columbia.edu/books/miseducation/.

Wysham, Daphne. 2012. "The Six Stages of Climate Grief." Other Words. September 3, 2012. https://otherwords.org/the_six_stages_of_climate_grief/.

Yeager, David S., and Gregory M. Walton. 2011. "Social-Psychological Interventions in Education: They're Not Magic." *Review of Educational Research* 81 (2): 267–301. https://doi.org/10.3102/0034654311405999.

3 Science, But Not Just Science

Whys and Wherefores of a Transdisciplinary Approach

3.1 An Argument for Transdisciplinarity in Climate Education

What Killed Stacy Ruffin?

During one week in the middle of August, 2016, seven trillion tons of water fell over a town in Louisiana. The resulting deluge killed thirteen people, including a 44-year-old woman called Stacy Ruffin. Stacy was returning from hospital with her mother in a truck driven by a neighbor when the truck was swept away in a flash flood. They had just suffered the loss of Stacy's brother in hospital. The water pulled all three of them out of the truck; Stacy's mother spent a harrowing 22 hours clinging to vegetation in water that was up to her armpits. But she survived, and so did the neighbor. Stacy didn't make it, leaving behind two children and a family mourning a devastating double loss.

What sets this tragedy apart from others that result from natural disasters is that this unusual rainstorm was not entirely a natural disaster. Climate scientists studying the relatively new field of weather attribution estimate that this rainstorm was made 40% more likely due to human-caused climate change. Which raises the question – who or what is to blame for Stacy's death?

This ethical question – and the interweaving of human tragedy with changes in the physical climate system – is brought out vividly in an article 'What Killed Stacy Ruffin?' by CNN reporter John C. Sutherland. Stacy Ruffin was an African American woman, holding a responsible position at a local Walmart, living in a trailer park with her family after the family home burned down. A loving mother and daughter, she was the go-to person in the family for all kinds of troubles. Interviewing Stacy's family as well as the scientists working on weather attribution, Sutherland presents us with a local, personal, viscerally real situation that has resulted, in part, from the relatively abstract problem of global climate change.

The three stories I have related so far, including the two in Chapter 1: The Scientist and the Elder, and The Village Women of Jharkhand – present lived reality for people in specific locales. Among the common threads is climate change. Once we see how climate change manifests – not as an

DOI: 10.4324/9781003294443-3

abstract global problem, but in *place*, affecting real people, its transdisciplinary nature becomes obvious. The story of Stacy Ruffin's tragic death and its
wider context includes multiple disciplinary threads: not only the science of
climate and weather, but also problems with urban planning and city design,
such as construction in flood-prone areas and lack of flood-proofing infrastructure; considerations of economics and politics; the health of local ecosystems; the history of African American impoverishment in the United States,
and questions of ethics and justice. While research in climate pedagogy is
still relatively scant, with major gaps in the Global South, there does seem
to be an emphasis on climate as a mostly scientific-technical subject (United
States Global Change Research Program 2009). We now know that treating
the complex problem of climate change in this narrow way may be counter-
productive – in fact, having knowledge about climate change does not necessarily inspire people to act, or act wisely with regard to the climate problem,
as I have described in Chapter 2.

> Climate change courses and curricular interventions primarily emphasize
> scientific literacy through a focus on physical processes, documentation of
> rising emissions, and empirical evidence of a changing climate. Classroom
> explorations of responses to climate change are often limited to "business-
> as-usual" policy options, new technologies, and behavioral interventions
> to reduce emissions or promote adaptation. As a result, students have
> difficulty recognizing social, psychological, and emotional dimensions of
> the issue, and often fail to see openings, possibilities, and entry points for
> active engagement with sustainability transformations.
>
> (Leichenko and O'Brien 2020)

In Chapter 2, I described how my early exclusive focus on climate science
gave rise to cognitive and affective disengagement for most students. That
learning about climate change can be emotionally difficult, and therefore the
emotional dimension needs to be acknowledged and addressed, is increasingly
evident from scholarly work, although this has not penetrated sufficiently into
actual teaching. My suspicion is that exclusively focusing on the science of
climate change may lead to students falling into traps such as 'climate doomism,' a state of despair that nothing can be done to fix the climate (Watts 2021)
leading to apathy, or the trap of the technofix (that technology will enable us to
innovate our way out of the crisis) or the market-fix (that the logic of market
forces will push us safely past disaster). An exclusive focus on the science
is likely to be less cognitively useful as a teaching tool, as the large body of
research on the efficacy of integrating socio-scientific issues in science education indicates (Zeidler, Herman, and Sadler 2019).

Ultimately, as I point out in my criteria for an effective pedagogy of climate
change in Chapter 2, if our purpose in educating students is to enable them
to become ethically motivated informed citizens inspired to take meaningful

action, then from that perspective also, a narrow focus on climate science is inadequate. How communities engage with and respond to the climate problem depends on their socio-cultural frameworks, their local historical contexts, and their identities within societies (Lee et al. 2020; Leichenko and O'Brien 2020). 'Mechanistic knowledge' of climate change, that is, knowledge of physical climatic processes and impacts, has been shown in some studies to interact in a complex way with worldview and quantitative reasoning among students (Zummo, Donovan, and Busch 2021). Referring to Kahan's application of the cultural theory of risk (Kahan et al. 2012), which presents worldview along two axes: individual to communitarian, and hierarchical to egalitarian, these scholars found that while mechanistic knowledge increased student receptivity to climate change in their sample, worldview was an influence, and quantitative reasoning skills amplified ideological polarization. Thus, information is not enough as a change agent. Even those who accept the validity of climate science and are concerned, need not be motivated to talk about global warming, let alone act – for example, the Yale Climate Communications 2022 survey on Climate Change in the American Mind indicates 64% who are concerned about climate change, and 67% who rarely, or never, talk about it (Leiserowitz, A. et al. 2022).

It is not surprising, then, that recent years have seen some significant work arguing for transdisciplinary approaches to climate education, crucially within the sciences (Singh 2021; Kubisch et al. 2022). Science literacy has also undergone an epistemological broadening; for instance, Science Literacy Vision III (Sjöström and Eilks 2018) and 'Critical scientific Literacy' (Hodson 2010) both recognize that science literacy must go beyond knowledge of the science, and even beyond the socio-cultural context of the science to sociopolitical action, and the recognition that science and technology often serve power, against the interests of the poor and marginalized. 'Youth are demanding action; science educators ought to enable learners and communities to transform and reinvent the world they are inheriting' (Kyle 2020). In the introduction to their book 'Teaching Climate Change in the United States,' editors Joseph Henderson and Andrea Dewes speak of an 'education that will shift social and material conditions in ways that lead to a tangible decline in carbon emissions and toward increased forms of resilience and flourishing for both humans and the more-than-human world,' and note that this will entail 'confronting entrenched systems of power.'

In their paper 'Broadening Epistemologies and Methodologies in Climate Change Education Research' (Busch, Henderson, and Stevenson 2019), the authors identify three paradigms for climate education research: positivist/post-positivist, social constructivist, and critical and transformative. This appears to mirror different paradigms of climate education in the classroom. The critical and transformative paradigm is not just about acquiring knowledge but also acting to confront power, seize agency, and make change. Given the urgency of the climate crisis and related social-environmental problems, it seems evident that transgressive and radical approaches (Lotz-Sisitka et al. 2015) are necessary. Certain climate scientists are openly advocating for climate action and taking part in protests (Quackenbush, Casey 2022), and among some climate scientists, the experience of engaging with communities for climate adaptation appears to be leading toward an epistemological awakening in climate science beyond the science itself (Shepherd and Lloyd 2021).

My contention is, therefore, that the impacts of climate change (ethical, societal, economic, and biophysical) and directions for meaningful action call for a transdisciplinary approach to climate change education informed by the best practices of transformational learning and transgressive learning. Here, I distinguish between multidisciplinarity (where multiple disciplines provide separate viewpoints on a particular subject), interdisciplinarity (in which two or more disciplines are combined in an integrative way), and transdisciplinarity (in which the distinction between disciplines is transcended to create a new way of thinking) (Leavy 2011). I acknowledge the difficulty of developing interdisciplinarity in the current siloed education system, let alone transdisciplinarity; in fact, many of the attempts toward a multifaceted approach to climate pedagogy stop at multidisciplinarity, where there is limited integration across disciplinary boundaries, especially between the natural sciences and the social sciences, and the natural sciences and humanities.

The 'how' of transdisciplinary work is also important to consider. In Chapter 5 and later, I will describe in some detail my attempts to apply a diffractive approach (Barad 2007) to transdisciplinarity, and the central role of stories in this regard. As scholars Vivienne Bozalek and Michalinos Zembylas state (Bozalek and Zembylas 2017):

> A diffractive methodology … is not setting up one approach/text/discipline against another but rather a detailed, attentive and careful reading the ideas of one through another, leading to more generative 'inventive provocations' … and the possibility of a true transdisciplinarity rather than interdisciplinarity.

3.2   Justice, Power, and the Four Dimensions

However, it is possible to have a transdisciplinary approach without making central the issues of justice and power. Therefore, it is necessary to specify that justice is, in fact, central to any serious consideration of the climate problem – rather than burying it in 'transdisciplinarity,' it must be front and center. Why?

The second common thread in the three stories I have referred to is precisely the centrality of justice and power. None of the communities represented by the Iñupiaq Elder of the Alaskan North Slope, the village women of Jharkhand, and Stacy Ruffin and her family are significant contributors to the climate crisis. Yet they, and those like them from other marginalized communities, are subject to its worst impacts. Indigenous people, the people of the Global South, the povertized, and the young bear the brunt of climate catastrophe despite not having created the problem. This is not to say that the privileged are immune from the climate crisis – the 2020 wildfires in California and the 2022 heat waves in Europe tell us quite clearly that no one is safe. However, the marginalized bear the greatest burden and risk. Moreover, as the story of the Village Women of Jharkhand illustrates, many people who are at the bottom of the pyramid of power have the courage, creativity, agency, and gumption to take meaningful action; yet their contributions are not recognized, and they do not have a seat at the policy table. Their perspectives and worldviews are at variance with those of modern industrial civilization; therefore, these alternative perspectives are sidelined, ignored, or sanitized. As I will show in later chapters, the climate crisis is therefore also an issue of epistemic injustice. Further details of how injustice and power hierarchies – and their connection to colonialism – are central to understanding climate change are elucidated in Chapter 5.

These stories illustrate the entangled nature of our social-environmental problems. As I will make clear in Chapter 4, climate change is not the only problem we face. In fact, carbon reductionism – an exclusive and usually technocentric focus on carbon dioxide as the driver of climate change – which dominates much of elite discourse today, is a dangerous trap, because we also need to pay attention to the wanton destruction of the biosphere that includes species mass extinctions, as well as less publicized issues like the imbalance in the nitrogen and phosphorus cycles. What becomes apparent through a deeper examination of stories like the Village Women of Jharkhand is that climate change is only one of a complex of problems that are all related at the root. This becomes crucial when considering purported solutions to the climate problem, as we shall see in Chapter 7.

I will mention briefly that the diffractive approach referred to earlier is necessarily inclusive of concerns of justice, equity, and power, since, as Karen Barad, physicist and feminist philosopher indicates (Barad 2007), ethics, epistemology, and ontology are inextricably entangled.

The above discussion points to the power of stories and case studies as portals and pathways across disciplinary silos. In Chapter 5, I will elaborate on the use of stories in the classroom as diffractive agents, transdisciplinary tools, and portal-pathways that enable journeys across apparent dichotomies – between science and humanities, between personal and political, between local and global, between human and Nature, and between paradigms, to name a few. Although I do not discuss case studies in this book, their use in education has its roots in law and medicine and case-based education now widely recognized as effective in multiple contexts ('Case-Based Learning' 2017). However, stories and case studies have to be carefully chosen to illustrate the transdisciplinary nature of the climate crisis and the centrality of justice. In addition, the context of the story, and audience frameworks that determine audience responses to it are crucial considerations.

Returning to the classroom, then, let me reiterate what I consider to be the four dimensions of an effective climate pedagogy, from Chapter 2. The Scientific-technological dimension explores the science, evidence, and impacts of global climate change and its local manifestations, with particular attention to their relationship. Often, the privileging of the global over the local reinforces top-down power dynamics that is not conducive to justice and ethics. Additionally, the justice issue enters this dimension through the question: why are these changes happening to the Earth's climate system? Who or what is responsible?

The transdisciplinary dimension – from my perspective as a physics educator – embraces the various disciplinary threads outside science that are essential for understanding the impacts of and responses to the climate problem among societies: history, including environmental history, economics, sociology, psychology, humanities, and the arts. Thus, the Iñupiaq experience of settler colonialism, the struggle for land rights since the 19th century, the shift in the relationship to the land with the passing of the Alaska Native Communities Settlement Act in 1971 that formed Native Corporations, the dilemmas with respect to oil drilling that confront the communities, the changing relationships with the land and the species on which the Iñupiat depend on for food as the Arctic sea ice melts – all these considerations enable us to understand more deeply what climate change feels to communities at ground zero. Justice and power are inherent in such an exploration.

The epistemological dimension – or more accurately, the onto-epistemological dimension – clarifies that we live within paradigms that shape and limit our imaginations, and – as I discuss in Chapter 4 and have hinted in Table 2.1 – the dominant paradigm of modern industrial civilization is at best inadequate for dealing with the climate crisis. Therefore, we need paradigms that embrace the key features or teachings of the problem itself, and as we shall see in Chapter 6, Indigenous onto-epistemologies provide inspiration for these.

Finally, the psychosocial action dimension needs to be brought out. How do we help students across psychological barriers like despair, anxiety, and apathy, so that they can inculcate an 'optimism of the will,' and not only find hope in the darkness but also make hope in the darkness? How do we make them aware of power hierarchies and give them the tools to help dismantle the pyramid of power? How might we help them feel their own power and agency? It is primarily for these reasons that traditional teaching approaches are not suitable for an effective pedagogy of climate change. If we are to demonstrate how power acts to destroy, to exploit, to corrupt, and co-opt, we must also turn the light on how power hierarchies manifest in our own spaces. Academia, sadly, tends to reflect rather than challenge the societal status quo. Thus, making the classroom as egalitarian a space as possible is crucial to teaching climate change effectively. For this, the role of teacher and student must be re-examined, which is why I suggest the power-flattening approaches inspired by transformative education in the previous chapter.

But how to bring these four dimensions into the classroom? As I mentioned when discussing institutional barriers to meaningful climate education in Chapter 2, the situation is less than ideal because what we really need is a complete dismantling of the education system to build something better. But while we work toward that end, we can still accomplish something in our classrooms, especially with some kind of 'teaching across disciplines' initiative where we can work with colleagues across the spectrum. Below, I describe how to plan an approach that educators can take in their own classrooms, based on my attempts to transform my teaching and my classroom, initially in almost complete isolation. It is my hope that my experience will give fellow educators, in formal and informal spaces, inspiration for their own transdisciplinary endeavors, eventually to bring needed change to their institutions and communities.

3.3   Course Planning: The Preliminary Work

Let us begin by considering how to integrate climate change into a pre-existing course that is not solely devoted to the subject – in my case, a general physics college course, but it could be anything from history to chemistry. Later, we will consider building an independent course on climate change that embraces these four dimensions, but most educators in schools and colleges are bound by a pre-existing curriculum, so the question of integrating climate change is paramount for those wishing to teach it in such a context.

The first step is to consider how the course and the discipline intersect with climate change. As an example, I show in Table 3.1 how topics in general physics courses (depth depending on the course) can align with key aspects of climate science. This is, of course, a piecemeal approach to teaching essential climate science, and my early attempts were failures in part because climate topics were scattered through the semester. I will discuss a little later how to develop a more integrative and holistic learning experience *despite* this problem.

We can see from Table 3.1 that a mere alignment of topics between the subject of the course and climate change does not embrace all four dimensions of the latter. But this is only the first step. Let us consider how this might look for someone teaching, for example, human biology (anatomy and physiology) and as a second example, a topic outside the sciences such as World History. These are assumed to be at the high school or beginning college level.

How might anatomy and physiology intersect with climate change? Human beings do not live in isolation – we are part of local ecosystems and the biosphere. Every breath we take involves an indebtedness to other lifeforms – trees and phytoplankton, across the planet, to begin with – and to processes over long timescales, such as the Great Oxidation Event some 2 billion years ago. Apart from respiration, the topic of body temperature regulation in mammals that maintains homeostasis allows for an exploration of the impact of heat waves on the human body, heat waves being a major impact of climate change. Heat illness ranges from mild symptoms to fatal heat strokes. Such questions

Table 3.1 Aligning physics topics with basic climate science

Physics Topics	Climate Change Topic
Contextualizing discussion (see next section for details) and introduction	Introduce Figure 3.1; elicit basic identification of Earth subsystems; define and distinguish climate and weather; discover the centrality of justice issues
Thermal energy, temperature	Blackbody radiation, Stefan–Boltzmann law, radiative equilibrium, equilibrium surface temperature of a planet; calculation for inner planets assuming no atmosphere; thermal expansion of ocean water and sea level rise
Fluids; flotation, laminar flow	Causes of sea level rise; sea ice and consequences of sea ice loss; ocean currents and their role in climate
Oscillations and resonance	Why carbon dioxide 'traps' infra-red radiation – resonance and the oscillation modes of CO_2
Waves and the electromagnetic spectrum	Connecting with earlier discussion on Blackbody radiation: Sun's radiation spectrum; natural greenhouse effect; Earth's radiation spectrum; actual surface temperatures for the inner planets; anthropogenic greenhouse effect; Earth's energy balance; simplified carbon cycle
Climate week	The Big Picture: connecting climate concepts via three meta-concepts: Balance/Imbalance, Critical Thresholds, Complex Interconnections
	Introduction to complex systems and connection to climate change
	Critical Thresholds and Planetary Boundaries
	Evidence, Impacts and climate justice
	What needs to be done?
	What are the barriers?
	Start interdisciplinary project (optional) that integrates science and other aspects of climate in a real world context
Electromagnetism	Oscillating dipoles and understanding more fully why CO_2 is a greenhouse gas
End of course	Optional project completed

as how the body maintains a healthy weight and sufficient energy for its needs (metabolic balance) are analogous to processes in the Earth system that generate energy balance and a consistent global average surface temperature. It is also well recognized (World Health Organization 2021) that climate change presents a major public health menace beyond the impacts of heat waves. The poleward movement of vector-borne tropical diseases, the potential awakening of pathogens currently dormant under ice sheets, the impact, physical and emotional, of extreme weather events all present challenges to the healthcare system. A 2021 report from the American Psychological Association, 'Mental Health and Our Changing Climate: Impacts, Inequities, Responses' (Susan Clayton et al. 2021) examines the unequally distributed impact of climate change on individuals and communities, from the direct impact of extreme weather to eco-anxiety among youth. 'Heat can fuel mood and anxiety disorders, schizophrenia, vascular dementia, use of emergency mental health services, suicide, interpersonal aggression, and violence. Drought can lead to stress, anxiety, depression, uncertainty, shame, humiliation, and suicide, particularly amongst farmers.' Issues of justice and power arise when we look at the disproportionate impact of climatic health hazards on marginalized groups, including lack of access to quality healthcare, and the prevalence of racist and colonialist structures in the mainstream medical system. The importance of Indigenous ways of knowing can illuminate the epistemological dimension, for example, the Swinomish tribe's radically holistic approach to health, that includes community and ecological health ('Swinomish Climate Change Initiative' n.d.). Thus, there are multiple intersection points between the subject of climate change and a course on human anatomy and physiology.

For a course outside the sciences, such as World History, for example, the instructor might find it useful to expand the four dimensions to five because from their perspective, 'transdisciplinary' would embrace all disciplines except history, and it would be natural to emphasize the historical dimension. However, even with such a course, the scientific-technological deserves a dimension all its own because a basic understanding of climate science is crucial to the understanding of the whole. I also wish to emphasize that these dimensions are all interrelated, and any walls between them are porous. For example, the history of developments in science and technology is intimately connected with historical forces such as conquest and colonialism. Therefore, an educator from the discipline of history interested in adapting this approach might consider the following five dimensions:

- The scientific-technological
- The historical
- The transdisciplinary
- The epistemological
- The psychosocial action

that also make central the issue of justice and power.

My colleague, historian Stefan Papaioannou, showed me a popular (in the United States) World History textbook, Worlds Together, Worlds Apart (Jeremy Adelman et al. 2021). The book already includes some references to climate change, in the context of how peoples and cultures are shaped by landscape, weather, climate and by changes therein, and how these drive movements, wars, and migration. It mentions the developing field of Big History, which is an attempt to include the large-scale story of the cosmos and expand history to the more-than-human world. (In my personal opinion, some proponents of Big History tend to go too far toward Grand Narratives, undermining local and alternative stories and epistemologies, but the field is still developing.) When I described my pedagogical framework to my colleague, he was optimistic that it could be of use to a teacher of history.

The task of determining intersections between climate change and one's course, is, of course, only the first step of a possible five.

1 Determine intersections between climate change and your coursework, including intersections with justice and ethics
2 Examine which of the four (or five) dimensions are covered through those topics, and which are 'left out'
3 Find ways to integrate those dimensions through both inter/transdisciplinary approaches (elaborated in the next few chapters) and by designating some course time for 'putting it all together' through Climate Week and an interdisciplinary project
4 Plan an experience that from the beginning establishes the transdisciplinary nature of the climate problem and the centrality of justice
5 Design a visual tool that will serve as a holistic conceptual device to deepen understanding and prevent fragmented or piecemeal learning

For a science educator, it is clear, after making a table such as Table 3.1 (through all but the last two rows) that all the dimensions except for 'scientific-technological' are left out, as are any considerations of justice and ethics. How to integrate these aspects into a science course? There are two paths that, in my experience, must be taken (both are important). One is to bring discussions of justice and ethics into discussions of science and technology, which may be considered an 'interdisciplinary approach' in the sense that I defined it above, and the other is to foreground the transdisciplinarity through stories and case studies. For an educator in the humanities or social sciences, the left-out aspects will include the scientific-technological dimension. Why and how might a teacher of history or literature introduce the key science of climate change? My argument is that first, it is necessary – just as it is necessary for a science educator to talk about justice and history – and second, it is possible, via three meta-concepts that I elaborate upon in Chapter 4. It is necessary because climate change is inherently transdisciplinary, and for students to have a holistic understanding, the disciplinary threads must be usefully intertwined, even as we naturally emphasize our discipline's perspective in the course. It is necessary also because the scientific speaks to the sociological and illuminates possible ways of engaging with the climate problem.

3.4 A Place to Begin: Introducing Students to the Climate Problem

For the remainder of this chapter, I will focus on items 4 and 5. I describe below the integrative experience I have designed for my students. This is introduced early in the semester as a stage-setting device. Before the semester begins, I email my students to invite them to the course, promising that the course will enable them to usefully explore an overarching question that is intriguing, important, and beautiful (Bain 2004). The question is about the nature of climate change and its impact on our world, and our futures. Then, within the first three weeks of the semester, we visit our university Planetarium (campuses without planetariums can alternatively use photographs as I detail below).

The Planetarium experience begins with a projection of Planet Earth seen from space, via an Earth science simulation software called Layered Earth (layeredearth.com n.d.). As the planet spins slowly, it looks like a blue-green jewel in the depths of space. I ask the students to examine it as if they were aliens watching the planet from a spaceship. What are its key features? There are oceans, and land. There are icy and snowy regions, and there is life, represented by the green of the Amazon and the Congo. Swirls of clouds indicate an atmosphere, barely visible at the edges of the planet as a thin veil. Students have discovered the five natural subsystems of the Earth's climate: the hydrosphere, lithosphere, cryosphere, biosphere and atmosphere. I hand out blank copies of a visual tool (Figure 3.1) so that students can fill in these five natural subsystems. Thus far in our exploration, there is no evidence of human presence or impact on the planet.

We then observe an image of Earth's night side. Now, human presence is vividly evident. Knots and streams of light form dense networks across the continents, except for Antarctica. Clearly, the presence of artificial light is an indicator of human impact on the planet. But wait! Why are some regions of the Earth so dark? Over the Amazon, and some parts of Africa and Asia, lights are few and far-between. I ask the students to hypothesize why this is so. Their typical first hypothesis is that the intensity is correlated with population density. We test this hypothesis by overlaying population density data over the Earth's image. We find, to our surprise, that while in some regions, darkness seems correlated to a low population density, such as over the Amazon, this is not true everywhere. Regions of Africa and Asia that are fairly densely populated are nevertheless relatively dark. Compared to these, North America and Europe are lit like a Christmas tree.

Now the students have to think more deeply. They come up with hypotheses to explain the dark, well-populated regions. Lack of access to technology, perhaps? But why would people in some areas not have access to technology? Then poverty comes up. Inequality. Colonialism. And it becomes immediately obvious that if we take the extent and degree of artificial light as a rough measure of human climatic impact on the planet (since we haven't yet transitioned away from fossil fuels to a sufficient degree), then not all humans and human social groups are equally responsible for the climate problem. We then identify 'marginalized peoples' as those who are generally denied power and agency in mainstream industrial cultures, including people from the Global South, people of color, women, Indigenous people, and the young.

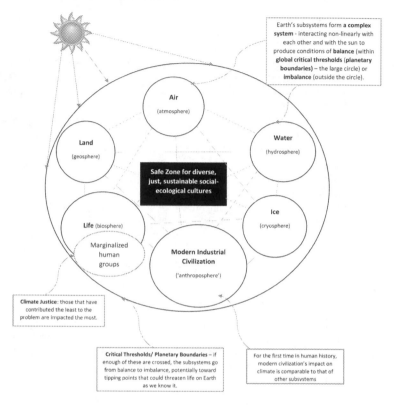

Figure 3.1 The subsystems of the Earth, including the three meta-concepts and the importance of justice.

As we go through this exercise in the dark of the planetarium, students are filling out the blank version of Figure 3.1 with the help of small, clip-on LED lights that give off a red glow. I introduce the term 'anthroposphere' as a sixth subsystem, proposed by some scientists and scholars to indicate that human activity on the planet has become a geological-scale influence on the Earth's climate. I ask if the term is fair, considering that there are some populations who have not contributed significantly to creating the climate problem. After some discussion, we settle on the alternative term 'modern industrial civiliza-tion' as being somewhat more accurate while acknowledging that under some circumstances, 'anthroposphere' remains useful. Thus, right from this prelimi-nary exercise at the start of the semester, we acknowledge the centrality of justice in our exploration of the climate problem.

This is the first time that Figure 3.1 is introduced. At this stage, we have filled out all the small circles within the large circle that symbolizes the Earth. I tell the students that each time we study a climate-related topic, we will return to this diagram, fill out more details, and develop a deeper understanding of the climate problem. Each time a phenomenon related to climate change is encountered, we will consider what subsystems of the climate system are involved. Thus, Figure 3.1 serves as a unifying visual tool that – along with the meta-concepts discussed in Chapter 4 – can help avoid a piecemeal, scattered exposure to climate topics throughout the semester.

It is possible for educators in other disciplines to construct their own, subject-specific holistic visual tools to accompany Figure 3.1, that serve this purpose.

References

Adelman, Jeremy, Elizabeth Pollard, Clifford Rosenberg, and Robert Tignor. 2021. *Worlds Together, Worlds Apart*. New York, NY: W.W. Norton. https://wwnorton.com/books/Worlds-Together-Worlds-Apart/.

Bain, Ken. 2004. *What the Best College Teachers Do*. Cambridge, MA: Harvard University Press.

Barad, Karen. 2007. *Meeting the Universe Halfway: Quantum Physics and the Entanglement of Matter and Meaning*. Durham, NC: Duke University Press.

Bozalek, Vivienne, and Michalinos Zembylas. 2017. "Diffraction or Reflection? Sketching the Contours of Two Methodologies in Educational Research." *International Journal of Qualitative Studies in Education* 30 (2): 111–27. https://doi.org/10.1080/09518398.2016.1201166.

Busch, K. C., Joseph A. Henderson, and Kathryn T. Stevenson. 2019. "Broadening Epistemologies and Methodologies in Climate Change Education Research." *Environmental Education Research* 25 (6): 955–71. https://doi.org/10.1080/13504622.2018.1514588.

"Case-Based Learning." 2017. Poorvu Center for Teaching and Learning. October 6, 2017. https://poorvucenter.yale.edu/strategic-resources-digital-publications/strategies-teaching/case-based-learning.

Clayton, Susan, Christie Manning, Meighen Speiser, and Alison Nicole Hill. 2021. "Mental Health and Our Changing Climate: Impacts, Inequities, Responses." American Psychological Association. https://www.apa.org/news/press/releases/mental-health-climate-change.pdf.

Hodson, Derek. 2010. "Science Education as a Call to Action." *Canadian Journal of Science, Mathematics and Technology Education* 10 (3): 197–206. https://doi.org/10.1080/14926156.2010.504478.

Kahan, Dan M., Ellen Peters, Maggie Wittlin, Paul Slovic, Lisa Larrimore Ouellette, Donald Braman, and Gregory Mandel. 2012. "The Polarizing Impact of Science Literacy and Numeracy on Perceived Climate Change Risks." *Nature Climate Change* 2 (10): 732–35. https://doi.org/10.1038/nclimate1547.

Kubisch, Susanne, Hanna Krimm, Nina Liebhaber, Karin Oberauer, Veronika Deisenrieder, Sandra Parth, Melanie Frick, Johann Stötter, and Lars Keller. 2022.

"Rethinking Quality Science Education for Climate Action: Transdisciplinary Education for Transformative Learning and Engagement." *Frontiers in Education* 7. https://www.frontiersin.org/articles/10.3389/feduc.2022.838135.

Kyle, William C. 2020. "Expanding Our Views of Science Education to Address Sustainable Development, Empowerment, and Social Transformation." *Disciplinary and Interdisciplinary Science Education Research* 2 (1): 2. https://doi.org/10.1186/s43031-019-0018-5.

layeredearth.com. n.d. "Layered Earth | Earth Science Simulation Software and Curriculum." Accessed July 21, 2023. https://www.layeredearth.com/.

Leavy. 2011. *Essentials of Transdisciplinary Research*. 1st ed. Abingdon-on-Thames, UK: Routledge.

Lee, Katharine, Nathalia Gjersoe, Saffron O'Neill, and Julie Barnett. 2020. "Youth Perceptions of Climate Change: A Narrative Synthesis." *WIREs Climate Change* 11 (3): e641. https://doi.org/10.1002/wcc.641.

Leichenko, Robin, and Karen O'Brien. 2020. "Teaching Climate Change in the Anthropocene: An Integrative Approach." *Anthropocene* 30 (June): 100241. https://doi.org/10.1016/j.ancene.2020.100241.

Leiserowitz, A., E. Maibach, S. Rosenthal, J. Kotcher, J. Carman, L. Neyens, and T. Myers, et al. 2022. "Climate Change in the American Mind, April 2022." Yale Program on Climate Change Communication. https://climatecommunication.yale.edu/publications/climate-change-in-the-american-mind-april-2022/.

Lotz-Sisitka, Heila, Arjen EJ Wals, David Kronlid, and Dylan McGarry. 2015. "Transformative, Transgressive Social Learning: Rethinking Higher Education Pedagogy in Times of Systemic Global Dysfunction." *Current Opinion in Environmental Sustainability*, Sustainability science, 16 (October): 73–80. https://doi.org/10.1016/j.cosust.2015.07.018.

Quackenbush, Casey. 2022. "The Climate Scientists Are Not Alright." *Washington Post*, May 24, 2022. https://www.washingtonpost.com/climate-environment/2022/05/20/climate-change-scientists-protests/.

Shepherd, Theodore G., and Elisabeth A. Lloyd. 2021. "Meaningful Climate Science." *Climatic Change* 169 (1): 17. https://doi.org/10.1007/s10584-021-03246-2.

Singh, Vandana. 2021. "Toward a Transdisciplinary, Justice-Centered Pedagogy of Climate Change." In *Curriculum and Learning for Climate Action: Toward an SDG 4.7 Roadmap for Systems Change*, edited by Radhika Iyengar and Christina Kwauk, 169–87. Leiden: Brill. https://doi.org/10.1163/9789004471818_010.

Sjöström, Jesper, and Ingo Eilks. 2018. "Reconsidering Different Visions of Scientific Literacy and Science Education Based on the Concept of Bildung." In *Cognition, Metacognition, and Culture in STEM Education: Learning, Teaching and Assessment*, edited by Yehudit Judy Dori, Zemira R. Mevarech, and Dale R. Baker, 65–88. Innovations in Science Education and Technology. Cham: Springer International Publishing. https://doi.org/10.1007/978-3-319-66659-4_4.

"Swinomish Climate Change Initiative." n.d. Swinomish Climate Change Initiative. Accessed April 2, 2023. https://www.swinomish-climate.com.

United States Global Change Research Program. 2009. "The Essential Principles of Climate Literacy | NOAA Climate.Gov." March 2009. https://www.climate.gov/teaching/essential-principles-climate-literacy/essential-principles-climate-literacy.

Watts, Jonathan. 2021. "Climatologist Michael E Mann: 'Good People Fall Victim to Doomism. I Do Too Sometimes.'" *The Observer*, February 27, 2021, sec. Environment.

https://www.theguardian.com/environment/2021/feb/27/climatologist-michael-e-mann-doomism-climate-crisis-interview.

World Health Organization. 2021. "Climate Change and Health." 2021. https://www.who.int/news-room/fact-sheets/detail/climate-change-and-health.

Zeidler, Dana L., Benjamin C. Herman, and Troy D. Sadler. 2019. "New Directions in Socioscientific Issues Research." *Disciplinary and Interdisciplinary Science Education Research* 1 (1): 11. https://doi.org/10.1186/s43031-019-0008-7.

Zummo, Lynne, Brian Donovan, and K. C. Busch. 2021. "Complex Influences of Mechanistic Knowledge, Worldview, and Quantitative Reasoning on Climate Change Discourse: Evidence for Ideologically Motivated Reasoning among Youth." *Journal of Research in Science Teaching* 58 (1): 95–127. https://doi.org/10.1002/tea.21648.

4   Science and More than Science
Three Transdisciplinary Meta-Concepts

4.1   Introduction

Listening to my explanation of the warming effect of carbon dioxide via the greenhouse effect, a student of mine asked, with bleak humor: 'So does that mean we shouldn't breathe out?' He went on to wonder about what would happen with an increasing population, all those mouths exhaling carbon dioxide into the atmosphere. It was only my second attempt at teaching climate change in one of my general physics courses, and I realized that I had omitted something of crucial importance. Through this and similar experiences, it dawned on me that simply teaching the properties of carbon dioxide and its effect on the Earth's temperature via the greenhouse effect – followed by a discussion of the evidence for and impacts of global warming – was woefully inadequate, as I have discussed in earlier chapters. Among other things, public misconceptions and misinformation about climate change, some of it likely fueled by climate deniers, was finding its way into my classroom. Confusing global heating with the ozone problem, mixing up climate and weather, suspecting climate scientists of being in it for the money – these were some of the instances I noted in the early years. These experiences made me wonder: what is it that *everyone* needs to know about the science of global climate change for a good understanding of the problem?

A meaningful understanding of climate change science for a non-expert would embrace its characteristic features, clarify confusions, inoculate against misinformation and disinformation, enable critical thinking about purported climate solutions, and inspire the right kind of action. Therefore, 'what everyone needs to know' cannot be a reductive list of 'essential' facts – it must be holistic, a sense-making *conceptual structure*, a scaffolding within which any new information about climate change can be emplaced and accommodated. It must provide the context for decision, judgment, and action.

In an attempt to answer the question: *what is it that everyone needs to know about the science of global climate change for a good understanding of the problem?* I have developed a conceptual structure that shows promise. Before we get into the details, however, it is important to do some preliminary stage-setting, following the planetarium experience I discussed in Chapter 3.

DOI: 10.4324/9781003294443-4

4.2 Preliminary Stage-Setting

The first item is to clarify the difference between climate and weather.

I begin by telling a little story of the harsh winter of 2015 in the Boston area. This was the last 'typical' Boston winter I experienced, with several heavy snowfalls. While I was shoveling some three or four feet of snow off my driveway, my neighbor was engaged in a similar effort. We paused for breath at about the same time, and he said to me angrily: 'so much for global warming, eh?'

It was not the right moment for a lecture on the difference between weather and climate, or the definition of 'global' in 'global warming,' but I am grateful to that neighbor for providing my students with a teaching moment. Science-resistant politicians have made similar statements. This little anecdote helps frame the following definitions.

Weather refers to the atmospheric conditions – temperature, precipitation, wind, etc. – over a period of a few days, across a small region. Thus, we speak of the weather in Boston or Delhi today or this week.

Climate, on the other hand, refers to atmospheric conditions over a large region and longer periods of time, typically decades to millennia. It is therefore the mean or average of weather, or more precisely, the statistics of weather, including its variability. We therefore speak of the humid continental climate of the northeastern United States where Boston is located, or the semi-arid climate of the Haryana-Rajasthan region where Delhi is located.

A common and helpful way to think about these concepts is this: weather is represented by what you wear when you go out today. Climate is represented by what you have in your closet.

Global climate refers to the climatic conditions averaged over the planet as a whole. The average global surface temperature of the Earth (about 14°C or 57°F) is a good measure of global climate, and indicates a rise of about 1.1°C above pre-industrial temperatures. This average is obtained from data from thousands of land weather stations around the world, along with ocean buoys, radar, and satellites. Thus, a severe winter in the Boston area doesn't mean that the Earth is not warming because the *global* average surface temperature obviously includes temperatures from all over the world. But it is also the case that such global averages cannot be directly felt or sensed, and for most people numbers such as '1.1°C rise since pre-industrial times' are meaningless without context, a point to which I will return later.

Having understood the meaning of global average surface temperature, we can ask the question: what factors influence this number? I then introduce the three dials for changing Earth's average surface temperature: first, the sun's activity. The sun is the main source of energy for life on Earth, and the intensity and nature of solar radiation falling on Earth is clearly going to influence climate on its attendant planets. Second, greenhouse gases: 'heat-trapping' gases like carbon dioxide and water vapor. Third, albedo, that is, the shininess or reflectivity of the Earth's surface. The icy and snowy parts of the Earth (the cryosphere) along with clouds, reflect without absorption some 30% of the sun's energy back into space, therefore limiting how much is retained in the Earth system. The last two dials are not necessarily independent of each other and often act in concert.

Which of these dials is responsible for current global heating? It turns out that the sun's activity is not to blame; it has a natural cycle of about 11 years, but there has been no net increase in the sun's activity since the 1950s, the very period in which we have seen unprecedented global heating. Nor is albedo (a combination of the effects of snow, ice, clouds, and aerosols) a dominant factor. There is clear and unambiguous evidence that the increase in greenhouse gases – carbon dioxide in particular – due to 'human activity' is the main reason why the Earth's average surface temperature is going up (IPCC 2023).

4.3   First Meta-Concept: Balance/Imbalance

Having set the stage, we can now introduce the conceptual framework. It consists of three large, transdisciplinary concepts I call meta-concepts: balance/imbalance, critical thresholds, and complex interconnections. These are not independent of each other, and we can start with any of them and discover the other two. Together, they form a conceptual scaffolding that enables a deeper and wider understanding of the basic science of the climate problem. In a later chapter (Chapter 6), I demonstrate the transdisciplinary nature of these meta-concepts.

Let us begin with the idea of Balance/Imbalance. We have an embodied sense of the terms *balance* and *imbalance*. If I stand on the toes of one foot, with my arms spread out, I can maintain balance for a few seconds. While I may sway back and forth a little, I have a certain resilience to falling over, that is, I have the capacity to recover my equilibrium, and on average I can state that I am in a state of balance. However, if somebody tugs on one arm while I am tilted their way, I might find myself falling over, unable to recover balance. As I fall, I am in a state of imbalance, going toward a new state of balance: being flat out on the floor.

It is instructive to begin a discussion of balance and imbalance in this embodied way, with students trying out various postures and attempting to freeze in them. They can then sense that there is a threshold beyond which balance is hard to maintain, beyond which one goes into imbalance. The next step is to illustrate *dynamic* balance and imbalance (or steady-state versus non-steady-state). For this, I employ an activity, thus.

I draw, with a marker, a large rectangle on the floor, with an entry point and an exit point (Figure 4.1). The size of the rectangle depends on the number of students and the dimensions of the classroom. I ask for a student volunteer to usher people into the rectangle, and another to usher people out of it. Other student volunteers cluster around the entry point of the rectangle. When the activity begins, I ask the person ushering in students to do so at the same speed

Figure 4.1 An activity to understand dynamic balance and imbalance.

as the person ushering out students. I ask students who enter the rectangle to wander around it a little, before allowing themselves to be ushered out at the far end. Meanwhile, a couple of students acting as observers mark points on a graph of 'number of students in the rectangle' versus time in (approximate) seconds.

At first, the number of students in the rectangle stays fairly steady, varying between four and five for a small class of 18 students. I then ask the person ushering in to speed up while the person ushering out slows down. It takes some practice, but when the students get this going, it becomes evident that the number of students within the rectangle starts to go up. If the ushers continue to speed up and slow down respectively, the number of students within the rectangle rises quite dramatically.

The graph shows this as well – initially, a more-or-less horizontal line, with small ups and downs, representing an average dynamic balance, followed by a steep rise, representing imbalance.

By this time, the students would have had their planetarium experience (described near the end of Chapter 3), and would have been introduced in a very basic way to the idea that carbon dioxide is a 'heat-trapping' gas. This exercise now allows me to introduce the carbon cycle in a highly simplified but still useful way. I explain that the rectangle represents the atmosphere; the person ushering in the students represents carbon sources, that is, all processes that add carbon dioxide to the atmosphere. The person ushering out students represents carbon sinks, that is, all processes that take away carbon dioxide from the atmosphere. We briefly list some carbon sources: respiration, volcanic eruptions, fossil fuel burning – and carbon sinks: photosynthesis by green plants, the absorption of carbon dioxide by the oceans. The students in the rectangle represent carbon dioxide molecules. When the carbon sources are acting at the same speed as carbon sinks, then the number of CO_2 molecules in the atmosphere is roughly constant, representing dynamic balance. But, as is currently the case, when carbon sources act faster than carbon sinks (as a result of increasing fossil fuel combustion and destruction of natural habitat such as forests), then atmospheric CO_2 concentration increases. In my anecdotal experience, most students and members of the public who have not had the benefit of a sound mathematical education have difficulty understanding rates of change. Thus, students will often describe balance/imbalance as having to do with the *amount* of carbon dioxide put in the atmosphere versus the amount pulled out rather than the *speeds* of these processes. The activity described above helps students understand the significance of rates of change, but the point needs to be emphasized multiple times.

When CO_2 concentration rises in the atmosphere, it is worth discussing how long the gas is expected to stay there. One of the diagrams examined by the students shows the carbon cycle operating over multiple timescales, the longest of which is the carbonate-silicate cycle with a time period of hundreds of thousands to a million years. Models indicate (Archer et al. 2009) that while about half the CO_2 humans release today will be taken up by the land biosphere and the surface oceans in a matter of a century or two, a significant fraction (20–35%) will persist in the atmosphere on a scale of centuries to millennia. The longevity of carbon dioxide in the atmosphere allows us to discuss the idea of intergenerational justice and the responsibility that those of us alive on Earth today have to several future generations of humans and nonhumans who will have to deal with the consequences of our actions. Zeroing anthropogenic carbon emissions would cause CO_2 concentrations to fall to a new, lower level within decades (Matthews and Weaver 2010) as we discuss in more detail when considering climate solutions.

Carbon cycle imbalance, as we have at present, leads inevitably to an imbalance in the Earth's energy budget. How and why this is so can help make the connection between rising CO_2 levels and rising temperature. I introduce this 'heat-trapping' property of carbon dioxide in a way that I hope is also useful and intriguing for teachers of non-science courses as well, although it may be omitted in these courses.

First, I introduce a fascinating truth about our universe; that everything in it can oscillate – tree branches, pendulums, bridges, molecules. The physics of oscillations would be part of most general physics courses, but even if it is not, the idea can be introduced in an interesting and playful manner. Poetically, we might say that everything in the universe dances – and each dance has its own particular rhythm, beat, or frequency. I suspend two pendulums of different lengths from a stand and let them swing in turn. As they swing, we clap each time the pendulum reaches one of its extreme points. The clapping rhythm is the beat of the pendulum. We notice that each pendulum, being a different system, has its own characteristic beat, with the longer one having the slower beat.

Now we are ready to understand the carbon dioxide molecule. It turns out that the carbon dioxide molecule can 'dance' or oscillate in three ways, each with its own beat or frequency or rhythm. The symmetric, asymmetric, and bending modes are illustrated in Figure 4.2. The bending mode has the

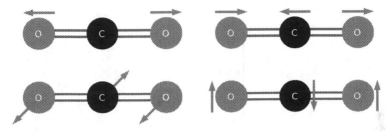

Figure 4.2 The oscillation modes of the carbon dioxide molecule. The top two diagrams show the symmetric and anti-symmetric modes. The diagrams at the bottom are two forms of the bending mode.

Credit: Markus Nielbock, Wikipedia.

slowest beat (lowest frequency) followed by the symmetric and the asymmetric modes. I teach the students these three dances of the carbon dioxide molecule, sometimes with a musical accompaniment; this application of embodied learning is not only enjoyable but also helps them remember the three modes of oscillation. The mode that is relevant for global heating is the bending mode.

One of the most beautiful phenomena in all of Nature is that of resonance. This under-appreciated yet ubiquitous phenomenon is responsible for the blue of the sky, the operation of a microwave oven, the sound of a musical instrument, and the heat-trapping effect of greenhouse gases like carbon dioxide. To understand it, I ask students to observe a swinging pendulum, and clap its beat, so that they have a sense of its natural or characteristic rhythm. Next, I 'interfere' with the pendulum by applying an external periodic force, or, to be less technical, simply tapping on it with a beat of my own. Initially, I deliberately tap on the pendulum with a beat much faster than its natural rhythm, and we see that the pendulum's swing becomes quickly erratic, and may even stop. Next, I tap on the pendulum with a beat that is similar to its own natural rhythm (that is, I match the frequency of my hand with the natural frequency of the pendulum), and a remarkable transformation occurs. The pendulum starts to swing with a wider and wider arc.

This phenomenon: the increase in the arc of swing (or the extent of oscillation) when the external rhythm matches the natural beat of the system is called resonance. Anyone who has ever pushed a child on a swing, responding to the child's demand 'swing me higher,' knows this. We naturally tune the frequency (or rhythm or beat) of our push to the natural frequency of the swing-plus-child.

Resonance can also be demonstrated with three pendulums suspended from the same stand. Two pendulums must have the same length, and the third should be very much longer or shorter. Let us then start one of the two identical pendulums swinging and watch what happens. The third pendulum might quiver a bit, but it is the identical twin that will start to swing on its own, in response to the swing of the first pendulum. The second pendulum resonates with the first *because they have the same natural rhythm.* This is, of course, the likely origin of the everyday expression 'I resonate with you!' Except that in physics, it is not about matching opinions, but matching rhythms (or frequencies or beats).

What has this got to do with global heating and Earth's energy balance? We are now in a position to understand – at a basic level – the essentials of the greenhouse effect. First, note that the term is misleading because the warming of the Earth has nothing to do with how greenhouses stay warm. Having stated that, let us consider Figure 4.3.

The sun is our main source of energy. It pours down on Earth waves of light, high-energy heat, and ultraviolet radiation. (A more advanced general physics class will go into details of these, but the details are not necessary for a good basic understanding.) The Earth's surface (unless it is icy or snowy) absorbs most of this energy and re-emits it in the form of low energy heat waves. These invisible heat waves coming off the Earth's surface happen to have the same frequency (or rhythm or beat) as that of the bending mode of a carbon dioxide molecule.

Imagine being a CO_2 molecule tumbling through the air. Perhaps you are not oscillating very much as you go your way. But then, suddenly, you are hit with a heat wave coming off the Earth, which has the same beat as your bending mode. What will you do? You will start to oscillate or dance in the bending mode! So, systems that have more than one mode of oscillation (or more than

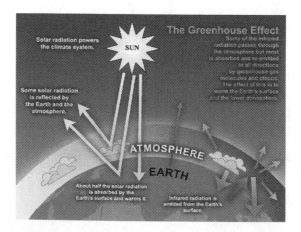

Figure 4.3 The greenhouse effect.
Graphic credit: IPCC (2007).

one way to dance) pick their mode of oscillation to match that of the external disturbance.

It turns out that it is in the bending mode that carbon dioxide is able to take in the energy of the outgoing heat wave and scatter it in all directions (this has to do with the generation of electromagnetic waves by oscillating dipoles, which can be elucidated in an advanced class but is not necessary for a basic understanding). Therefore, a heat wave that would have gone out into the cold emptiness of space is now being bounced around in all directions in the atmosphere by the carbon dioxide molecule. Some of it goes sideways, some of it goes back toward the ground, and some of it goes into space.

The result is remarkable. To appreciate it, it is important to point out that carbon dioxide is a trace gas in the atmosphere, barely 0.04% of all the gases (which are mainly nitrogen and oxygen). Its concentration is so miniscule that it is measured in parts per million. Historically, before the industrial revolution, carbon dioxide levels in the atmosphere did not rise above 300 parts per million – that is, there were approximately 300 particles of CO_2 out of every million particles of our atmosphere (most of which would be nitrogen and oxygen) –300 ppm was the upper limit for at least the past million years. And yet, this tiny amount of CO_2 has raised the Earth's temperature through this 'heat-trapping' process from a chilly −19°C to a bearable +14°C, allowing the flourishing of life as we know it on Earth! The natural greenhouse effect is therefore essential to life as we know it. It is evident that a little CO_2 goes a long way.

In physics and chemistry courses, the greenhouse effect can be explored more deeply via the Stefan-Boltzmann equation. The best aid for educators in these disciplines that I've found is David Archer's very useful and lucid book, Global Warming: Understanding the Forecast (Archer, David 2011), and the series of short video lectures associated with it, freely available on the University of Chicago website (Archer, David n.d.). Also on the website are links to toy climate models that allow students to play with greenhouse gas and solar radiation dynamics.

We can now see that balance in the Earth's carbon cycle implies (in the absence of other changes) balance in the Earth's energy budget, via the above-described greenhouse effect. Balance in the carbon cycle means that carbon sources act as fast as carbon sinks, maintaining a roughly constant amount of CO_2 in the atmosphere. Balance in Earth's energy means that incoming solar radiation (excluding what's reflected) comes into the Earth system as fast as the Earth radiates out energy in the form of low-energy heat waves. This would mean that a roughly constant amount of energy is retained in the Earth system, leading to a more-or-less constant average global surface temperature.

Therefore, an imbalance in the carbon cycle resulting in increasing carbon dioxide levels in the atmosphere will, through the greenhouse effect, cause an imbalance in Earth's energy budget. Solar radiation is coming in as quickly as it did before, but the extra CO_2 molecules in the atmosphere are causing the outgoing energy to leave Earth at a slower rate. The result? Energy build-up in the Earth system, leading to rising temperatures. The Earth's energy imbalance is also affected by vanishing sea ice and decreasing cloud cover, as we'll see when we get to Chapter 6 to consider the crucial importance of the cryosphere in Earth's climate.

After this experience, three historical graphs of CO_2 concentration in the atmosphere (two of which are reproduced below, Figures 4.4 and 4.5) help us recognize – and complicate – our hitherto somewhat simple conceptions of balance and imbalance and also give us context for the current crisis of climate change. There is rich interdisciplinary information in these graphs, including the role of human societies.

The graph below shows carbon dioxide concentrations in the atmosphere from nearly a million years ago to recent times. The first striking observation is that *change* is the norm. Carbon dioxide levels do not stay the same through the millennia, but fluctuate up and down in what must be natural cycles since humans appear on this graph only between 200,000 and 300,000 years before

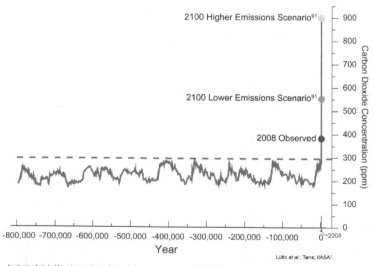

Analysis of air bubbles trapped in an Antarctic ice core extending back 800,000 years documents the Earth's changing carbon dioxide concentration. Over this long period, natural factors have caused the atmospheric carbon dioxide concentration to vary within a range of about 170 to 300 parts per million (ppm). Temperature-related data make clear that these variations have played a central role in determining the global climate. As a result of human activities, the present carbon dioxide concentration of about 385 ppm is about 30 percent above its highest level over at least the last 800,000 years. In the absence of strong control measures, emissions projected for this century would result in the carbon dioxide concentration increasing to a level that is roughly 2 to 3 times the highest level occurring over the glacial-interglacial era that spans the last 800,000 or more years.

Figure 4.4 Changes in carbon dioxide concentrations in the atmosphere over the past 800,000 years.

Image Source: Globalchange.gov 2009, Archived.

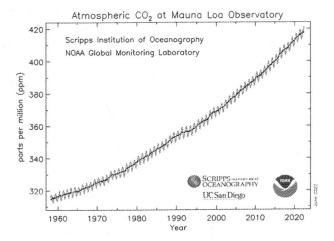

Figure 4.5 Carbon dioxide concentration in the atmosphere, more recent.
Credit: NOAA.

today, and the Industrial revolution is only 150 years or so from the right-hand end of the graph. Constant carbon dioxide levels (balance) would be indicated by flat (horizontal) lines, and at first sight there aren't any. However, depending on the timescale in which we are interested, we can find regions where carbon dioxide is roughly constant, which implies averaging over fluctuations within that timescale. See, for instance, the region of the graph just before the 400,000 year mark. Thus, the concept of dynamic balance is a convenient and helpful construct that *depends on the time-scale of interest.*

The graph shows that carbon dioxide levels in Earth's atmosphere have changed between 200 and 300 ppm, but have not gone above 300 ppm until recently. The changes in carbon dioxide levels by just 100 ppm represent dramatic changes in Earth's climate. These natural cycles are thought to be the result of variations in the Earth's orbit due to the gravitational influences of the other planets in our solar system: the Milankovitch cycles that result in alternating glacial and interglacial periods. Dips in CO_2 concentration coincide with glacial periods, when glaciers covered much of the northern hemisphere, whereas the peaks coincide with interglacial warm periods such as the one in which we are living today, the Holocene. Where the graph is flat, we have on average a roughly constant amount of carbon dioxide in the atmosphere averaged over that time period, representing balance. Where the graph slopes, carbon dioxide levels are changing, implying imbalance. It is useful to have students draw circles on the graph for examples of each of these conditions. It is worth emphasizing the importance of timescale – for instance an expansion of the time axis in the above graph will reveal that the sloping regions also have periods where carbon dioxide levels did not change very much.

The graph tells us that both balance and imbalance have occurred naturally in the Earth system well before humans appeared on the scene, which helps to dispel the popular misconception that Nature is always seeking, or in, a state of 'balance.' The scale of the first graph does not allow us to see the details of the rise in carbon dioxide concentrations as we approach the era of the Industrial revolution, but the spike is dramatic. It tells us that current levels of atmospheric carbon dioxide (415 ppm in 2022) are unprecedented in nearly a million years of Earth's climate history. It is also apparent that the cycle of glacial-interglacial periods is likely broken. It further projects future carbon dioxide levels (for 2100) based on two future scenarios.

The next graph I use is Figure 2a from Shakun et al. (2012), behind a paywall (but accessible on the Internet here (John Cook 2023)). It is a close-up of the first one, focusing on the period between 22,000 and 7,000 years before the present day. We see here the correlation between carbon dioxide and temperature, as well as the emergence from the last glacial period to the current interglacial warm period, the Holocene epoch, that began some 12,000 years ago. It is this period, following the retreat of the glaciers to the far north, and the prevalence of relatively warm surface temperatures, that has allowed humanity to develop in all its splendid diversity, leading to our current globalized industrial civilization. Again, the horizontal portions of the graph represent periods of carbon cycle and energy balance (averaged over the centuries) while the sloping regions represent change or imbalance.

The third graph, Figure 4.5, shows us how globalized industrial civilization has impacted our planet through increasing carbon dioxide emissions. This is the famous Keeling curve, a close-up of the spike in Figure 4.4, showing us the rise in carbon dioxide levels since about 1960. We see, also, the ups and downs of seasonal carbon dioxide changes – since much of Earth's land mass is in the Northern hemisphere, the loss of foliage during winters results in a rise in carbon dioxide levels, followed by a drop when trees begin to leaf in the spring. But the upward trend is clear.

The third graph clarifies the imbalance represented by the spike in the first graph. It should be clear that the extent of the imbalance is extraordinary. With the onset of the industrial revolution, we have seen a steep rise in CO_2 levels in the past 150 years, with no possibility of return within the timescale of the glacial-interglacial cycle (Archer et al. 2009). It is useful also to consider the rate at which atmospheric CO_2 levels are rising (over 2 ppm/year in the last few years); in the past 60 years, the rate of increase is 100 times greater than the rate of change at the end of the last glacial era, 11,000 to 17,000 years ago (Lindsey, R. 2020). The sharp rise in both CO_2 and temperature since the industrial revolution points to the burning of fossil fuels as a *likely* factor, but without additional information (such as the known warming effects of CO_2), it establishes a correlation and not a cause. This is a good place to discuss how science actually works – evidence is built from multiple sources that converge, and cause is established via well-tested experiments, observations, and models that relate the variables in a manner consistent with physical laws. Time permitting, different kinds of uncertainty and confidence levels as applied to climate change (the epistemology of the field is still developing) may also be discussed in higher level classes (Risbey and O'Kane 2011; Lewis, S. and Gallant, A. 2013; IPCC 2014), including direct evidence of the 'human footprint' through isotope data. For most classrooms, however, what I have discussed above is enough to support the following as the evidence that global heating is real and human-caused:

- Carbon dioxide is a 'heat-trapping' gas, this property is well known (since 1850s) and can be reproduced in any well-equipped physical chemistry lab (field trip)
- When fossil fuels are burned, they emit carbon dioxide gas (easily established)
- Paleoclimate and instrumental records indicate a rise in carbon dioxide levels since the Industrial revolution, which is when fossil fuels began to be burned on a mass scale

Note that in my discussion of the terms Balance and Imbalance, it is evident that these terms have multiple meanings. Static balance, as exemplified by a person standing on the toes of one leg, is not the same as dynamic balance as indicated by my simplified carbon cycle model. And, as my subsequent discussion of critical thresholds and complex systems indicates, a sufficiently complex system such as an ecosystem can have multiple *regimes of stability*, a feature which can further enrich our understanding of Balance/Imbalance. Through these considerations, terms like Balance/Imbalance that are fuzzy in common use can become sharp and distinct within specific contexts; this can be made apparent to students (see also the discussion on boundary objects in Chapters 5 and 6).

4.4 Second Meta-Concept: Critical Thresholds

What happens when a system in balance goes into imbalance? Let us consider my earlier example of static balance: standing on the toes of one foot, with my arms spread out. As I sway back and forth, within a certain degree of tilt, I can recover my equilibrium and stay in balance. However, let's say that while I'm tilted to my left, somebody tugs on my left arm. I now tilt further than I would otherwise and cross an invisible boundary that commits me to change in one direction: I start to fall.

Thus, going from balance to imbalance implies crossing a critical threshold or boundary, after which I can no longer go back to my old state, at least within a certain timeframe. Such critical thresholds occur in both human and non-human contexts. Ecosystems, for example, can go from states of relative stability to abrupt change that is irreversible over those timescales. It is important to note here that educators must avoid some conceptual traps in talking about transitions in ecosystems. The notion of ecosystem balance is no longer in common use among ecologists; ecosystems are dynamic entities that are always changing, from both internal processes and external disturbances over multiple time-scales; stasis rarely occurs. Therefore, ecologists prefer to talk about ecosystems having 'multiple basins of attraction' and focus on resilience models instead of equilibrium models with which to understand responses to disturbances (Folke 2006). (A dynamical system possesses 'attractors,' which are states toward which the system tends to evolve through time, if left to itself; we can think of an attractor metaphorically as a valley in a mountainous topography, where water tends to collect from different sources along the watershed; a basin of attraction is the set of all starting states that end up at the particular attractor.) Ecosystems are also complicated by internal imperatives such as natural selection, which is why it is easier to talk about balance in the carbon cycle than balance in ecosystems. Further, the notion of the 'balance of Nature' is a widespread misconception that has led to problematic public and policy perceptions – assumptions that Nature 'seeks' balance, that balance is natural and imbalance is always due to the interference of humans. As we saw from the graphs above, imbalances in the carbon cycle have occurred throughout Earth's long history, without any assistance from humans. Therefore, when we consider critical thresholds at the local scale, one should either avoid the term 'balance' altogether or specify that there are multiple possible stability regimes defined in a contextual, place-based, time-scale-dependent manner such that short-term stable stages are viewed always against the backdrop of longer-term changes. For instance, we might consider a forest ecosystem in India upon which a village community depends and has done so for a few 100 years where changes are slow enough that on average there is balance. Or we might be interested in the coastal Arctic ecosystem of the Alaskan North Slope, on which the Iñupiat have depended for at least 4,000 years. Within such timeframes, there would likely be changes in the numbers and distributions of species, from internal dynamics, gradual evolutionary changes due to natural selection as well as larger scale external factors. In fact, Indigenous communities that have survived and thrived for millennia in their traditional lands have not done so by staying still, but by changing with the changes through a mutual relationship with their environment. But there are multiple scales and degrees of change. Therefore, 'balance' in this context refers to the resilience of a system in remaining within a particular state for a particular period of time, in contrast to a subsequent, dramatic change leading toward a new state of relative stability.

Critical transitions in Nature abound (Marten Scheffer 2009). These include the dramatic regime shift from abundant vegetation in North Africa to the Sahara desert, the classic transition from kelp forests to urchin barrens when sea otters were overhunted, and the change of state of Jamaican coral reefs following the wiping out of sea urchins due to a pathogen. Another example is the consequences of the recent introduction of wolves in Yellowstone National Park (Sustainable Human 2014). Human societies can also go from periods of relative stability to radical change.

Critical thresholds are transition regions in the evolution of a system, where it shifts from a state that (on a specified timescale) might be one of rough balance or stability, toward another, possibly very different future state. Such thresholds exist and are experienced at local scales – for example, consider the changes underway as Arctic sea ice melts. In some (not all) regions, polar bears are losing weight (Whiteman 2018) as their main prey, seals, become fewer. Polar bears have been observed eating birds' eggs and entering towns in search of food (Kamala Kelkar 2016; Sofia Moutinho 2021). The lack of seals and the disappearance of sea ice are also generating major changes in Iñupiaq ways of living, already undergoing larger-scale changes dating from the history of colonialism, the struggle for Native land rights, and the impact of modernity and the oil industry. If the latter, slower changes are on the scale of decades, it is possible that the shifts in some cultural practices, such as the deep relationship to the sea ice and dependence on seals and bowhead whales as fresh winter food, might become faster as the Arctic warms at four times the global rate.

Critical thresholds also occur at the global scale. The still-developing concept of planetary boundaries (Steffen et al. 2015) in this context is both interesting and pedagogically useful, although it is not without deserved critiques and responses (Biermann and Kim 2020; Brand et al. 2021). Planetary boundaries are thresholds defined by certain natural processes, cycles or tolerances, and so far, nine have been identified. These are biosphere integrity, climate change, biogeochemical cycles, land-system change, novel entities, stratospheric ozone depletion, ocean acidification, freshwater use and the global hydrological cycle, and atmospheric aerosol loading. Violating even one planetary boundary can shift the Earth system from a relatively stable state defined by certain characteristics toward a potential new steady state in the future (Figure 4.6).

Not all such new states of balance may be conducive to modern industrial civilization, or even to the existence of humans, so our discussion of planetary boundaries is biased toward conditions that are optimized for human well-being (but not necessarily modern industrial civilization as it is today). (It is to be noted that there is no one idea of 'human well-being,' therefore, it is important to invite pluralistic conceptualizations of the term (Kothari, A. et al. 2019).)

Crossing any one of these boundaries can imperil the conditions that make life (as we know it) possible; we have already crossed six of the nine in 2023. Note that a recent paper (Steffen et al. 2018) considers the Earth system to be at a critical point, a fork in the road past the glacial-interglacial cycles,

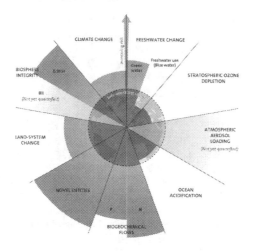

Figure 4.6 The nine planetary boundaries.

Source: Wikipedia. Designed by Azote for Stockholm Resilience Center.

where future action or inaction will determine whether we end up with a human-nurtured stabilized Earth or one that careens uncontrollably toward a hothouse state not seen for millions of years of Earth history.

One pedagogically useful aspect of the planetary boundaries concept is the fact that it emphasizes our indebtedness to natural cycles and processes. It reminds us that we are dependent on the larger nonhuman biophysical context for our survival, a fundamental truth conveniently ignored by modern industrial civilization and much of mainstream economics. A second important point is that the planetary boundaries concept demonstrates that climate change is not the sole social-environmental problem facing the biosphere. Any inter-to-transdisciplinary approach to teaching climate change should contextualize it within larger issues. Discussion of planetary boundaries is an opportunity to briefly introduce other violations of these boundaries – the imbalance in the nitrogen cycle, for example, as well as the dire situation of biodiversity loss: a million species threatened with extinction, as reported in a recent UN publication (Intergovernmental Science-Policy Platform on Biodiversity and Ecosystem Services 2019). This allows students to appreciate that climate change is part of a much larger complex of problems, and therefore technological 'solutions' for climate change alone that ignore the larger, deeper issues are unlikely to be helpful in the long term. These two implications of the planetary boundaries concept – that there are limits to human exploitation of the planet that we violate at our peril, and that we are confronted not with just the problem of climate change, but, in fact, a *polycrisis* – invites a critical exploration of the underlying socio-economic system and power hierarchy that is the common root of these interrelated crises.

The idea of critical thresholds as applied specifically to climate change is introduced via this question: what is the 'safe upper limit' of CO_2 concentration in the atmosphere? Or 'what is the "safe upper limit" for the Earth's global average surface temperature?' In 2009, governments and scientists agreed at the UN Copenhagen meeting that 2°C was the ceiling beyond which global temperature could not be allowed to rise. This was controversial, as there is no cut-and-dried formula for 'safe upper limit,' the definition of which term is as much political as it is scientific. In Paris in 2015, the aspirational goal was set to 1.5°C, (Gao, Gao, and Zhang 2017) although the world is likely to be heading toward a greater than 4°C rise by 2100 if no action is taken (Tollefson, J. 2020). However, some important research in 2011 (Meinshausen et al. 2009; Leaton, J. 2011) estimated that only 565 Gigatons of CO_2 could be permitted into the atmosphere before reaching a 2°C rise, whereas we already have enough global fossil fuel reserves that, if burned, would release about five times this 'safe' amount. Currently, fossil fuel companies are spending billions of dollars looking for new and increasingly hard-to-extract deposits of fossil fuels, despite the fact that current reserves are already in excess of climate limits (Carbon Tracker Initiative 2017; Grant, A. and Coffin, M. 2019). Seen within this metaconceptual framework, such 'business-as-usual' activities stand out as senseless and dangerous.

It is helpful to introduce the key findings of the 2018 IPCC Special report comparing a 1.5°C rise with a 2°C rise (IPCC 2018). Given that our current 1°C rise already presents us with substantial threats to human and natural systems, a 1.5°C world, which will be challenging, is far preferable to a 2°C world, which would likely be catastrophic.

A critique of the planetary boundaries concept – currently being debated, adapted and contested. – is necessary despite its usefulness as elaborated above. It is important to even briefly acknowledge that the concept of planetary boundaries and the related notion of a 'safe operating space for humanity' are not a purely scientifically determinable but are entangled with social norms, values, and processes; therefore, a need to democratize the concept has been articulated (Pickering and Persson 2020). This, along with the question – *What processes are causing us to cross planetary boundaries, and who benefits from these processes?* gives us the opportunity to revisit climate justice and its entanglement with each of the meta-concepts. Currently, I am moving toward the phrasing: 'diverse, just, sustainable social-ecological cultures' rather than 'safe operating space for humanity.' I am also in the process of integrating a study (O'Neill et al. 2018) that indicates that 'no country meets basic needs for its citizens at a globally sustainable level of resource use,' which sobering fact indicts the socio-cultural-economic system of modern industrial civilization.

I note that in 2023, some of the original researchers on planetary boundaries, along with collaborators from the social sciences, have significantly revised the planetary boundaries framework to include issues of justice and human well-being (Rockström et al. 2023). The proposed safe *and* just Earth System Boundaries framework quantifies eight such boundaries. I will be examining this proposal and its potential application in the classroom in future work.

4.5   Third Meta-Concept: Complex Interconnections

Some questions that have arisen in my classroom include:

a   How can humans possibly affect something as large as the Earth?
b   If climate change is really happening, why not wait to fix it until, for example, the climate of Boston is like that of North Carolina?
c   Technology is going to fix the problem!
d   Can we predict when we are about to cross a critical threshold? Can we reverse the breaking of a planetary boundary?

Directly or indirectly, these questions are about the interaction between human social systems (especially globalized modern industrial civilization) and the biophysical systems upon which we depend. To attempt to answer these, we must understand the nature of these systems. In classical physics, a system is typically an object or objects in whose behavior we are interested. Everything outside the system is its environment. Thus, the definition of a system really depends on the question we are asking because the question determines the boundary that delineates the system from its environment.

In much of classical physics, the systems we study are simple systems. That is, the system consists of clearly defined parts, each of which has a fixed function or role. A great metaphor for a simple system is a clock. The springs and gears that constitute the clock each have a specific role to play. The clock is entirely controllable and predictable. Because we can understand the clock by focusing on the role and behavior of each part, the clock is amenable to reductionism. In fact, science as we know owes a lot to reductionism, the idea that to understand the whole of a system, we need to divide it into parts and study the parts and their roles.

However, there are many real-world systems that are not fully amenable to reductionism; for these the Aristotelian holistic dictum 'the whole is greater than the sum of its parts' applies. Complex systems include the human nervous system, the human endocrine system, ecosystems, social networks, global financial networks, and global climate. While working physicists are well aware of the limitations of unbridled reductionism, and often employ what might be called 'optimal reductionism' (Hari Dass, N.D. 2019), the dominance of reductionist methods in science has meant that the detailed study of complex systems as *systems* (rather than solely as a conglomeration of parts) is fairly recent.

(Here, I am not being reductionist about reductionism and holism – we in fact need both). Currently, no widely agreed-upon definition of complexity exists. I align with Paul Fieguth's description of complex systems as large, dynamical non-linear, non-Gaussian, coupled spatial systems (Fieguth, P. 2017). Complex systems are made up of parts that interact non-linearly in time, in ways that make it difficult to predict details of their behavior compared to simple systems. Some (not all) complex systems exhibit chaotic behavior – for example, weather is chaotic but climate is not (Annan, J. and Connolley, W. 2005). Complex systems exhibit hysteresis – that is, having reached a certain state, they cannot be made to return to a previous state simply by reversing the conditions. Because they are so much easier to study, most systems that students encounter are small, linear and Gaussian. Therefore, introducing complexity, especially in the context of climate change in a physics classroom, is challenging.

Having introduced a simple system by way of an example like the clock – which may be complicated, but is not complex – I then introduce complex systems as *systems in which the interactions are as important as the parts* so that merely understanding the parts and how they function cannot lead us to a classical understanding (including predictability of details) of the whole. Complex systems are characterized by an inherent relationality between the parts, so much so, that a shift in relationships can change the nature and function of the parts. It is also the case that how the parts of a system are identified depends on the question being asked – the same system might be considered to be constituted differently for a different question. An example of a complex system is the human endocrine system, where a knowledge of how each gland works is insufficient to predict the exact behavior of the system as a whole. It is important to emphasize that complex systems, while not predictable in the same way that the trajectory of a projectile is predictable, do not behave arbitrarily – we can glean important information from them, but it is not always useful to ask the same questions of a complex system as we do of simple systems.

Once the idea of a complex system is introduced and a couple of examples presented, students often begin to recognize the presence of such systems everywhere around them. This is because most real-world systems are complex; the conceptual scaffolding afforded by our brief foray into complexity allows students for the first time to make sense of hitherto unacknowledged everyday presences of complex systems. A lab or demonstration that contrasts the behavior of a simple pendulum with that of the chaotic pendulum is also instructive in making this clear to students (although it is important to point out that not all complex systems are chaotic).

In our discussion, we point out the ways in which complex systems differ from simple systems. For the purposes of studying climate change, the following characteristics are important:

a   The existence of multiple, interacting feedback loops, both stabilizing (negative) (such as temperature regulation in the human body, or carbon uptake by the solid Earth on timescales of over 100,000 years) and destabilizing (positive) such as the ice-albedo feedback in the climate system (see, for instance, this NASA lesson (NASA 2019); feedback loops exist in simple systems as well, but complex systems may possess multiple feedback loops (many of which could be positive) that can interact with each other at different scales with varying levels of complexity; further, it is possible for feedback loops to flip from negative to positive, as in ocean absorption/emission of CO_2 and

b Tipping Elements: aspects of the climate system that, once certain thresholds are crossed, result in the system being committed to change in a particular direction, although the timing of that change can vary from abrupt to gradual. This includes for example the melting of the ice sheets. Such changes are generally irreversible on human timescales (Lenton et al. 2008; Lenton, Timothy M. et al. 2019).

c Tipping elements imply teleconnections: relationships between widely separated climate and weather phenomena. Teleconnections demonstrate the spatial interconnectivity of biophysical phenomena across the planet (see, e.g., Liu et al. 2023).

This intense relationality between regional climate and weather systems can cause cascading, systemic changes in the Earth's climate as a whole (Steffen et al. 2018).

The discussion on global climate as a complex system is begun by referring again to our unifying visual tool from Chapter 3, reproduced below, and considering the Earth's sub-systems as one way (among others) of subdividing the Earth into parts. The lines from the sun to the Earth, and those connecting each bubble, represent interactions, which, I remind students, are strong, nonlinear, and changing with time (e.g., by revisiting the fact that small changes in insolation due to orbital variations are connected with changes in ocean currents and atmospheric CO_2 levels that together give rise to the dramatic phenomenon of glacial-interglacial cycles as in Figure 4.4). Students then study graphs and animations of the sea ice extent in the Arctic. The ice is melting because the average surface temperature in the Arctic is rising at four times the global rate. Why might that be happening? This is a topic of ongoing research, and while the quantification of various feedbacks in the Arctic is an active area of inquiry (Goosse et al. 2018; Stuecker et al. 2018), the Arctic provides some of the most pedagogically useful examples of positive feedback loops in the climate system. These include the ice-albedo feedback, the permafrost-methane feedback, and what I call the Arctic Drilling feedback loop. These are discussed in more detail in Chapter 6, where I explain how the two 'natural' feedback loops can interact with each other and with the Arctic drilling feedback loop, potentially accelerating warming. This dramatic illustration of nonlinearity helps answer the question of how an initially small effect can become very large very quickly. We also briefly discuss how

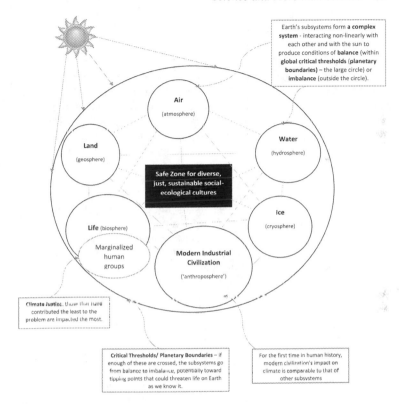

Earth's subsystems form **a complex system** - interacting non-linearly with each other and with the sun to produce conditions of **balance** (within **global critical thresholds (planetary boundaries)** – the large circle) or **imbalance** (outside the circle).

Air

(atmosphere)

Water

(hydrosphere)

Land

(geosphere)

Safe Zone for diverse, just, sustainable social-ecological cultures

Ice

(cryosphere)

Life (biosphere)

Marginalized human groups

Modern Industrial Civilization

('anthroposphere')

Climate Justice, those that have contributed the least to the problem are impacted the most.

Critical Thresholds/ Planetary Boundaries – if enough of these are crossed, the subsystems go from balance to imbalance, potentially toward tipping points that could threaten life on Earth as we know it.

For the first time in human history, modern civilization's impact on climate is comparable to that of other subsystems

Figure 4.7 A unifying visual tool.

stabilizing feedback loops (such as ocean absorption of carbon dioxide, or the role of tropical forests in sequestering carbon) can become destabilizing under warmer conditions. The issue of tipping points potentially leading to global tipping cascades is an active area of research (Lenton, Timothy M. et al. 2019) – passing several tipping points is likely to lead to irreversible, runaway or catastrophic climate change, after which human intervention is not likely to stabilize or reverse it (Figure 4.7).

In our brief foray into complex systems science, students are often astonished that they have not encountered these concepts in their prior educational experience. This is usually after they have come to understand and recognize the ubiquity of complex systems in the real world, an experience that can be revelatory. How is it that such an important aspect of the world has been denied them? This provides an opportunity for some important reflections on the history of science and the idea of paradigm shifts. I find the term 'paradigm' pedagogically useful despite the banality and ubiquity of its current usage. I begin with 'paradigm' in the sense of a 'disciplinary matrix' elucidated in a later work of Thomas Kuhn (Second Thoughts on Paradigms), as a collection of practices, symbolic generalizations, models, and exemplars that make and are made by a scientific community (Kuhn, Thomas 2012; Bird 2018). I extend this notion of a scientific paradigm to encompass social groups beyond the scientific community which, even if they are not aware of the details of the science, support, benefit from, are affected by and promulgate a worldview built (accurately or not) from this disciplinary matrix so that it becomes the scaffolding for a mostly unquestioned consensus reality. This is, therefore, a socio-scientific paradigm. Although there are many critiques of Kuhn's formulation of the progress of science via 'normal science' and 'revolutions,' I find some utility in the sociological-scientific extension of the idea of a paradigm – and paradigm shifts – as formulated by Kuhn. I acknowledge that this extension is an oversimplification of complex historical and socio-cultural processes, disputes, and debates; however, inspired by Shapin (Shapin, Steven 2018), I aver that following certain historical threads in order to understand how we got to this moment in human history can be a worthwhile endeavor. Using the term paradigm in this broader sense, I attempt to trace the development of certain ways of thinking and conceptualizing that are prevalent and dominant today.

In the classroom I motivate this discussion through a brief exploration of the shift from geocentrism to heliocentrism, and the changes wrought in both the cosmological and sociological realms. I follow this with a discussion of ancient to modern conceptualizations of the atom. These topics are already in the syllabus of the course Physics, Nature, and Society. I then introduce what may be called the Newtonian paradigm (admittedly somewhat unfair to Newton the person), also referred to as the Mechanistic or 'clockwork universe.' The term 'Newtonian Paradigm' encompasses the mechanical philosophies of Boyle, Kepler, Mersenne, and Descartes among others, and their mathematization via Newton's laws, as well as the atomic materialism of Boyle and Descartes (Boas 1952; Derek J. de Solla Price 1964; Shapin, Steven 2018). This amalgam gives rise to a view of the universe that is reductionist, mechanistic, and deterministic (Heylighen, F., Cilliers, P., and Gershenson, C. 2007). The analog clock is immediately seen as a metaphor of the Newtonian paradigm. In order to give students some idea of the historical complexity that accompanies the development of large ideas, we briefly discuss objections to the mechanistic universe (e.g., that of William Blake (Moore, A. 2014)).

We first critique the Newtonian Paradigm from the physics perspective. While Newtonian physics is powerful, we know that it has a limited domain of validity. It fails in the realm of the very small, very fast, and very massive. I introduce Figure 4.8. The x-y plane is inspired by very similar diagrams that have appeared in general physics textbooks (French, A.P. 1971; Hobson, Art 2009). Note that the various conceptual realms do not have hard boundaries,

A non-Newtonian Universe

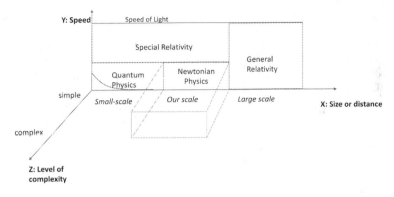

Figure 4.8 The domain of validity of Newtonian Physics.

and there is considerable overlap between them; thus, Newtonian physics can be considered an approximation of quantum physics, and from the other direction, of relativity. My addition of the z-axis acknowledges a revolution in physics and beyond, which is the physics of complexly entangled systems at all scales. The gradation from simple to complex is qualitative, as we don't have universal measures of complexity, and the designation of a system as 'complex' depends in part on the question we are interested in asking since the same system, when considered from different angles, may be simple or complex. Figure 4.8 allows students to understand the damage done when we think within the Newtonian box about systems that exist outside it – for example, some people in the medical profession still refer to the human body as a machine, when in fact the mechanistic model of the body is severely limited (as any endocrinologist or neurologist would agree). Even for a complex nonlinear system like global climate, the mechanistic view can be encountered in popular thinking (e.g., a 2009 planetarium show about Earth systems and climate change, Dynamic Earth (E&S Shows 2009) refers to the Earth explicitly as 'a machine which is the sum of its parts'). Yet the clockwork, mechanistic analog is a poor one for a system in which the interactions between the parts is so significant that it can change the nature of the parts – where oceans or forests might become, for example, net emitters rather than absorbers of carbon.

The power of this diagram is that it makes the invisible visible – our conceptual frameworks through which we make sense of the world suddenly become apparent, and unexamined or default assumptions become susceptible to scrutiny. Classroom debates and discussions based on this diagram often center on the possibility that science itself changes and evolves, and that a paradigm shift (in the broader sense) may well be taking place at this moment in history. The extent to which old ways of thinking persist despite changes in scientific understanding is also an interesting topic of discussion that complicates the idea of a paradigm shift. For instance, we may point to influences of Einstein's theory of relativity in philosophy and art, but the way humans live and think in modern industrial societies is mostly unaffected (except for the use of technological innovations owed to relativity) by this seismic shift in theoretical physics. There are many factors that prevent scientific revolutions from changing the way people at large conceptualize the world but among these is the role of power structures in society. Who benefits from the status quo? How do systems of power control how people conceptualize the world? While a physics class is not adequate for answering such questions in any depth, simply raising them makes for interesting speculative discussions, and prepares us for thinking about climate justice. Students also look at the pros and cons of large-scale technofixes (geoengineering schemes such as the injection of aerosols in the upper atmosphere to reflect away sunlight) (Oxford Geoengineering Programme 2018) via this metacognitive view of science – are such geoengineering schemes an example of Newtonian thinking about a non-Newtonian problem? I contrast the key features of the climate problem with the Newtonian paradigm, as summarized in Table 2.1, to emphasize the dramatic mismatch between the two, and to introduce the term *paradigm blindness* – the notion that the dominant socio-scientific paradigm in which one is immersed blinds us to imagine, comprehend, or consider seriously an alternative worldview that has a different underlying conceptual structure. The realization that a genuinely sustainable world involves not only technological and societal change but also a major worldview shift is revelatory for many students. Recent work on how different societies, especially the now globalized neoliberal culture, construe human-Nature relationships within a Dominant Social Paradigm aligns with this approach (Bogert et al. 2022). I discuss paradigms in the context of climate solutions in more detail in Chapter 7.

The idea that complex systems science is an essential component of effective climate change pedagogy is supported by other researchers (Roychoudhury, A., Shepardson, D.P., and Hirsch, A.S. 2017) and by the US government's Climate Literacy Guide https://www.climate.gov/teaching/climate but often, the discussion tends to be ahistorical and limited to the physical sciences. To take full advantage of the conceptual revelations offered by the subject, it is important that the teaching practice incorporate systems thinking methodology throughout – for example I use concept maps extensively, as well as small-group work, stock-and-flow analogies such as the Carbon Bathtub, and discussion of real-world dilemmas and situations through stories. However, far more work needs to be done to develop a truly systems-based pedagogy in the context of climate change.

For example, a potentially problematic feature of 'systems thinking' that I have observed in multiple settings is the assumed separation of the thinker from the system, as though the person observing the system is outside it. This is a false dichotomy that encourages a mindset of separatedness. We must always ask the question: where am I in this system? What is my role and that of my community, and how does it shift and change? Where am I in the hierarchy of power? An additional problem arises if we define the system narrowly and ignore the boundaries where our system of interest connects with other, larger systems. For example, an economist will typically consider the local or global economic system, while ignoring the natural biophysical systems within which it is embedded.

Therefore, much needs to be done to improve our understanding and teaching of complex systems. But that we must understand and teach these is not in doubt. Educators might be tempted to teach what is simple because it is simpler to teach, assuming that complex systems are too complex for students, especially young students to understand; however, this is not borne out by research (Grotzer, Tina A., 2012). This tendency to oversimplify what is not simple also has serious implications for policy (Stirling 2010), all the more reason to integrate the study of complex systems into education. My own attempts to bring to students' awareness the existence and importance of complex systems do not go far enough. Although students react with fascination, curiosity, and, in several cases, a desire to learn more, I find it impossible to do full justice to this subject in a traditional physics course based on Newtonian physics.

Ultimately, the very idea of complex systems is an artifact of the Newtonian Paradigm. If we accept complex relationality as an inherent feature of the universe, that is, the notion that simple systems are the exception rather than the rule, there would be no need for the term 'complex system' and the Newtonian Paradigm would remain safely within its domain of validity, without the wide, transdisciplinary influence it enjoys today. As we'll see in Chapter 6, Indigenous epistemologies tend to take a more sophisticated view of the centrality and ubiquity of complex systems.

4.6 A Brief Note on the Three Meta-Concepts and Transdisciplinarity

In what sense are the three meta-concepts transdisciplinary? As I elaborate in Chapters 5 and 6, these meta-concepts are broad enough that they have meaning in other disciplines, even if the connotation and implications are different from their usage in the basic science of climate change. Balance/Imbalance, for example, can be interpreted as stability versus change, concepts that are prevalent in disciplines from history to literature. Similarly, critical thresholds appear in multiple contexts, from daily life to every discipline that comes to mind. If complex relationality is inherent in the world – social and biophysical – then it must have relevance in every field of study. Thus, the three meta-concepts are, at the very least, interdisciplinary. Transdisciplinarity, however, implies a transcending of boundaries in order to create new knowledge, as I've mentioned in Chapter 3. Stories allow us to begin with the world and only then discover the disciplinary threads: the subject of Chapters 5 and 6.

My point here is not that there should be something universal and/or fixed about these three meta-concepts, or that there should be any effort to unify them across disciplines. Instead, their specific meanings within a discipline and the differences and similarities in meaning across disciplines can allow us to travel across, and ultimately transcend boundaries. Just as two water waves in a pond or kitchen sink superimpose to create patterns different from the original waves, the similarities *and* differences that these meta-concepts evoke across disciplines allow new questions, ideas, and concepts to emerge. This is the heart of a diffractive approach to transdisciplinarity, as I illustrate in Chapters 5, 6, and 10 in more detail.

References

Annan, J., and W. Connolley 2005. "Chaos and Climate." *RealClimate* (blog). September 2005. http://www.realclimate.org/index.php/archives/2005/11/chaos-and-climate/.

Archer, David. 2011. *Global Warming: Understanding the Forecast, 2nd Edition* | *Wiley*. Hoboken, NJ: Wiley. https://www.wiley.com/en-us/Global+Warming%3A+Understanding+the+Forecast%2C+2nd+Edition-p-9780470943410.

———. n.d. "'Global Warming: Understanding the Forecast' by David Archer." Video Lectures. Accessed April 15, 2023. https://forecast.uchicago.edu/.

Archer, David, Michael Eby, Victor Brovkin, Andy Ridgwell, Long Cao, Uwe Mikolajewicz, and Ken Caldeira, et al. 2009. "Atmospheric Lifetime of Fossil Fuel Carbon Dioxide." *Annual Review of Earth and Planetary Sciences* 37 (1): 117–34. https://doi.org/10.1146/annurev.earth.031208.100206.

Biermann, Frank, and Rakhyun E. Kim. 2020. "The Boundaries of the Planetary Boundary Framework: A Critical Appraisal of Approaches to Define a 'Safe Operating Space' for Humanity." *Annual Review of Environment and Resources* 45 (1): 497–521. https://doi.org/10.1146/annurev-environ-012320-080337.

Bird, Alexander. 2018. "Thomas Kuhn." In *The Stanford Encyclopedia of Philosophy*, edited by Edward N. Zalta, Winter 2018. Stanford: Metaphysics Research Lab, Stanford University. https://plato.stanford.edu/archives/win2018/entrieshomas-kuhn/.

Boas, Marie. 1952. "The Establishment of the Mechanical Philosophy." *Osiris* 10: 412–541. https://doi.org/10.1086/368562.

Bogert, Jeanne, Jacintha Ellers, Stephan Lewandowsky, Meena Balgopal, and Jeffrey Harvey. 2022. "Reviewing the Relationship between Neoliberal Societies and Nature: Implications of the Industrialized Dominant Social Paradigm for a Sustainable Future." *Ecology and Society* 27 (2). https://doi.org/10.5751/ES-13134-270207.

Brand, Ulrich, Barbara Muraca, Éric Pineault, Marlyne Sahakian, Anke Schaffartzik, Andreas Novy, and Christoph Streissler, et al. 2021. "From Planetary to Societal Boundaries: An Argument for Collectively Defined Self-Limitation." *Sustainability: Science, Practice and Policy* 17 (1): 265–92. https://doi.org/10.1080/15487733.2021.1940754.

Carbon Tracker Initiative. 2017. "Carbon Bubble." Carbon Tracker Initiative. August 2017. https://carbontracker.org/terms/carbon-bubble/.

Cook, John. 2023. "CO_2 Lags Temperature – What Does It Mean?" Skeptical Science. 2023. https://skepticalscience.com/argument.php?p=8&t=634&&a=7.

de Solla Price, Derek J. 1964. "Automata and the Origins of Mechanism and Mechanistic Philosophy." *Technology and Culture* 5 (1): 9–23. https://doi.org/10.2307/3101119.

E&S Shows. 2009. "Dynamic Earth – E&S Digital Theater Show." 2009. http://es.com/Shows/dynamicearth.

Fieguth, P. 2017. *An Introduction to Complex Systems: Society, Ecology and Nonlinear Dynamics.* New York, NY: Springer.

Folke, Carl. 2006. "Resilience: The Emergence of a Perspective for Social–Ecological Systems Analyses." *Global Environmental Change*, Resilience, Vulnerability, and Adaptation: A Cross-Cutting Theme of the International Human Dimensions Programme on Global Environmental Change, 16 (3). 253–67. https://doi.org/10.1016/j.gloenvcha.2006.04.002.

French, A.P. 1971. *Newtonian Mechanics.* First. MIT Introductory Physics. New York, NY: W.W. Norton.

Gao, Yun, Xiang Gao, and Xiaohua Zhang. 2017. "The 2°C Global Temperature Target and the Evolution of the Long-Term Goal of Addressing Climate Change—From the United Nations Framework Convention on Climate Change to the Paris Agreement." *Engineering* 3 (2): 272–78. https://doi.org/10.1016/J.ENG.2017.01.022.

Goosse, Hugues, Jennifer E. Kay, Kyle C. Armour, Alejandro Dodus-Salcedo, Helene Chepfer, David Docquier, and Alexandra Jonko, et al. 2018. "Quantifying Climate Feedbacks in Polar Regions." *Nature Communications* 9 (1): 1919. https://doi.org/10.1038/s41467-018-04173-0.

Grant, A., and M. Coffin 2019. "Breaking the Habit: Why None of the Large Oil Companies Are 'Paris Aligned' and What They Need to Do to Get There." Carbon Tracker Initiative. https://carbontracker.org/reports/breaking-the-habit/.

Grotzer, Tina A. 2012. *Learning Causality in a Complex World: Understandings of Consequence.* Washington, DC: Rowman and Littlefield. https://rowman.com/ISBN/9781610488631/Learning-Causality-in-a-Complex-World-Understandings-of-Consequence.

Hari Dass, N.D. 2019. "My Climate Pedagogy Article," September 15, 2019.

Heylighen, F., P. Cilliers, and C. Gershenson 2007. "Complexity and Philosophy." In *Complexity, Science and Society*, edited by Bogg, Jan and Geyer, Robert, First, 184. Abingdon-on-Thames, UK: Routledge. https://arxiv.org/ftp/cs/papers/0604/0604072.pdf.

Hobson, Art. 2009. *Physics: Concepts and Connections.* Fifth ed. New York, NY: Pearson.

Intergovernmental Science-Policy Platform on Biodiversity and Ecosystem Services. 2019. "Global Assessment Report on Biodiversity and Ecosystem Services." Bonn, Germany: IPBES Secretariat.

IPCC. 2007. "What Is the Greenhouse Effect?" FAQ 1.3 – AR4 WGI Chapter 1: Historical Overview of Climate Change Science. 2007. https://archive.ipcc.ch/publications_and_data/ar4/wg1/en/faq-1-3.html.

———. 2014. "IPCC, 2014: Climate Change 2014: Synthesis Report. Contribution of Working Groups I, II and III to the Fifth Assessment Report of the Intergovernmental Panel on Climate Change." Geneva, Switzerland: IPCC.

———. 2018. "Global Warming of 1.5°C: An IPCC Special Report on the Impacts of Global Warming of 1.5°C above Pre-Industrial Levels and Related Global Greenhouse Gas Emission Pathways, in the Context of Strengthening the Global Response to the Threat of Climate Change, Sustainable Development, and Efforts to Eradicate Poverty." https://www.ipcc.ch/sr15/.

———. 2023. "Synthesis Report of the IPCC Sixth Assessment Report (AR6)." https://www.ipcc.ch/report/sixth-assessment-report-cycle/.

Kelkar, Kamala. 2016. "Polar Bears, Growing Desperate for Food, Threaten Alaska Natives." PBS NewsHour. October 15, 2016. https://www.pbs.org/newshour/nation/polar-bears-food-native-alaskans.

Kothari, A., A. Salleh, A. Escobar, F. Demaria, and A. Acosta, eds. 2019. *Pluriverse: A Post-Development Dictionary*. New Delhi: Tulika Books.

Kuhn, Thomas. 2012. *The Structure of Scientific Revolutions*. Fourth. Chicago, IL: University of Chicago Press.

Leaton, J. 2011. "Unburnable Carbon: Are the World's Financial Markets Carrying a Carbon Bubble?" Carbon Tracker Initiative. file:///C:/Users/vsingh/Downloads/Unburnable-Carbon-Full-rev2-1.pdf.

Lenton, Timothy M., Hermann Held, Elmar Kriegler, Jim W. Hall, Wolfgang Lucht, Stefan Rahmstorf, and Hans Joachim Schellnhuber. 2008. "Tipping Elements in the Earth's Climate System." *Proceedings of the National Academy of Sciences* 105 (6): 1786–93. https://doi.org/10.1073/pnas.0705414105.

Lenton, Timothy M., Johan Rockström, Owen Gaffney, Stefan Rahmstorf, Katherine Richardson, Will Steffen, and Hans J. Schellnhuber 2019. "Climate Tipping Points – Too Risky to Bet Against." *Nature* 575 (November): 592–95. https://doi.org/doi:10.1038/d41586-019-03595-0.

Lewis, S., and A. Gallant 2013. "In Science, the Only Certainty Is Uncertainty." The Conversation. August 22, 2013. https://theconversation.com/in-science-the-only-certainty-is-uncertainty-17180.

Lindsey, R. 2020. "Climate Change: Atmospheric Carbon Dioxide." Climate.Gov: Science and Information for a Climate-Smart Nation. February 2020. https://www.climate.gov/news-features/understanding-climate/climate-change-atmospheric-carbon-dioxide.

Liu, Teng, Dean Chen, Lan Yang, Jun Meng, Zanchenling Wang, Josef Ludescher, and Jingfang Fan, et al. 2023. "Teleconnections Among Tipping Elements in the Earth System." *Nature Climate Change* 13 (1): 67–74. https://doi.org/10.1038/s41558-022-01558-4.

Matthews, H. Damon, and Andrew J. Weaver. 2010. "Committed Climate Warming." *Nature Geoscience* 3 (3): 142–43. https://doi.org/10.1038/ngeo813.

Meinshausen, Malte, Nicolai Meinshausen, William Hare, Sarah C. B. Raper, Katja Frieler, Reto Knutti, David J. Frame, and Myles R. Allen. 2009. "Greenhouse-Gas Emission Targets for Limiting Global Warming to 2°C." *Nature* 458 (7242): 1158–62. https://doi.org/10.1038/nature08017.

Moore, A. 2014. "Alan Moore on William Blake's Contempt for Newton | Blog | Royal Academy of Arts." December 2014. https://www.royalacademy.org.uk/article/william-blake-isaac-newton-ashmolean-oxford.

Moutinho, Sofia. 2021. "Hungry Polar Bears Are Struggling to Hunt Seabird Eggs." News from Science. April 7, 2021. https://www.science.org/content/article/hungry-polar-bears-are-struggling-hunt-seabird-eggs.

NASA. 2019. "Positive Feedback – Arctic Albedo." Lesson Plans. My NASA Data. My NASA Data. September 23, 2019. https://mynasadata.larc.nasa.gov/lesson-plans/positive-feedback-arctic-albedo.

O'Neill, Daniel W., Andrew L. Fanning, William F. Lamb, and Julia K. Steinberger. 2018. "A Good Life for All Within Planetary Boundaries." *Nature Sustainability* 1 (2): 88–95. https://doi.org/10.1038/s41893-018-0021-4.

Oxford Geoengineering Programme. 2018. "What Is Geoengineering?" 2018. http://www.geoengineering.ox.ac.uk/www.geoengineering.ox.ac.uk/what-is-geoengineering/what-is-geoengineering/index.html.

Pickering, Jonathan, and Åsa Persson. 2020. "Democratising Planetary Boundaries: Experts, Social Values and Deliberative Risk Evaluation in Earth System Governance." *Journal of Environmental Policy & Planning* 22 (1): 59–71. https://doi.org/10.1080/1523908X.2019.1661233.

Risbey, James S., and Terence J. O'Kane. 2011. "Sources of Knowledge and Ignorance in Climate Research." *Climatic Change* 108 (4): 755. https://doi.org/10.1007/s10584-011-0186-6.

Rockström, Johan, Joyeeta Gupta, Dahe Qin, Steven J. Lade, Jesse F. Abrams, Lauren S. Andersen, and David I. Armstrong McKay, et al. 2023. "Safe and Just Earth System Boundaries." *Nature*, May, 1–10. https://doi.org/10.1038/s41586-023-06083-8.

Roychoudhury, A., D.P. Shepardson, and A.S. Hirsch 2017. "System Thinking and Teaching in the Context of Climate System and Climate Change." In *Teaching and Learning About Climate Change: A Framework for Educators*, First. Abingdon-on-Thames: Routledge.

Scheffer, Marten. 2009. *Critical Transitions in Nature and Society | Princeton University Press*. Princeton, NJ: Princeton University Press. https://press.princeton.edu/books/paperback/9780691122045/critical-transitions-in-nature-and-society.

Shakun, Jeremy D., Peter U. Clark, Feng He, Shaun A. Marcott, Alan C. Mix, Zhengyu Liu, Bette Otto-Bliesner, Andreas Schmittner, and Edouard Bard. 2012. "Global Warming Preceded by Increasing Carbon Dioxide Concentrations During the Last Deglaciation." *Nature* 484 (7392): 49–54. https://doi.org/10.1038/nature10915.

Shapin, Steven. 2018. *The Scientific Revolution*. Second ed. Chicago, IL: University of Chicago Press.

Steffen, Will, Katherine Richardson, Johan Rockström, Sarah E. Cornell, Ingo Fetzer, Elena M. Bennett, and Reinette Biggs, et al. 2015. "Planetary Boundaries: Guiding Human Development on a Changing Planet." *Science* 347 (6223). https://doi.org/10.1126/science.1259855.

Steffen, Will, Johan Rockström, Katherine Richardson, Timothy M. Lenton, Carl Folke, Diana Liverman, and Colin P. Summerhayes, et al. 2018. "Trajectories of the Earth System in the Anthropocene." *Proceedings of the National Academy of Sciences* 115 (33): 8252–59. https://doi.org/10.1073/pnas.1810141115.

Stirling, Andy. 2010. "Keep It Complex." *Nature* 468 (7327): 1029–31. https://doi.org/10.1038/4681029a.

Stuecker, Malte F., Cecilia M. Bitz, Kyle C. Armour, Cristian Proistosescu, Sarah M. Kang, Shang-Ping Xie, and Doyeon Kim, et al. 2018. "Polar Amplification Dominated by Local Forcing and Feedbacks." *Nature Climate Change* 8 (12): 1076–81. https://doi.org/10.1038/s41558-018-0339-y.

Sustainable Human, dir. 2014. *How Wolves Change Rivers*. https://www.youtube.com/watch?v=ysa5OBhXz-Q.
Tollefson, J. 2020. "How Hot Will Earth Get by 2100?" *Nature* 580 (April): 443–45. https://doi.org/10.1038/d41586-020-01125-x.
Whiteman, John P. 2018. "Out of Balance in the Arctic." *Science* 359 (6375): 514–15. https://doi.org/10.1126/science.aar6723.

5 The Power of Stories

Foregrounding Justice in the (Science) Classroom

5.1 Stories Illustrate Injustice, Inequity, and Power Hierarchies

In this chapter, I first examine how these stories used in this framework enable us to embrace all four dimensions of an effective pedagogy of climate change, and how they allow us to center issues of justice and power. I will then go on to discuss other key roles that stories play in the classroom.

Let me begin with a traditional tale of the Iñupiat, as retold by Edna Ahgeak MacLean, Elder, educator and linguist (Chance 2002).

One day an avingaq decided to venture outside his hole and assess the rest of the world. When he stood up on his hind legs, lo and behold, to his surprise, he was able to reach the heavens. When he reached down, he felt the ground. When he reached in all directions, he was able to touch the limits of the world. He concluded that he was the largest person on the face of the earth.

In reality, the poor mouse had surfaced from his hole in the ground into an old Iñupiaq boot sole turned upside down. The top of his heaven was the sole of the atungak and the outer limits of his world were the sides of the atungak.

What does this story teach us? More generally, what role do stories play in the classroom, in particular, a college science classroom? This juxtaposition may seem incongruous, but I have found it invaluable for an effective pedagogy of climate change.

In a freshman seminar on Arctic climate change, for example, I begin with the story of The Scientist and the Elder that I described in Chapter 1, and follow that up with the above story, which I call The Mouse on the Tundra. In my general physics classes, I typically introduce stories such as The Scientist and the Elder, The Village Women and the Forest, and What Killed Stacy Ruffin. As I described in Chapter 1, some typical student questions about these stories include:

1   How did the Elder know the ice was going to break?
2   Why did the ice break?

DOI: 10.4324/9781003294443-5

3   What does The Village Women and the Forest story have to do with climate change?
4   How did poor village women with no education manage to restore a forest?

In Chapter 3, I discussed how the transdisciplinarity of the climate problem can be revealed and explored through story. Student questions like the ones above naturally lead to transdisciplinary excursions in the sciences, in Indigenous cultures and knowledge systems, in economics and history. This is because my suite of stories, real-world and fictionalized, are themselves transdisciplinary. After all, the real world doesn't draw hard lines between physics, poetry, art, and politics – these co-exist, interpenetrate, and entangle with each other. It is for convenience that we divide the world into disciplines, and while the distinctions are important, it is equally important not to take them too seriously in certain real-world contexts. Consider sea ice: not sea, not land, it exists in a liminal zone, transcending the physics and chemistry of its formation, the biology of the lifewebs it supports through its intricate inner structure, the place it occupies in the hearts of the Iñupiat as their metaphor of home and means of survival. Its role in the Earth's glacial-interglacial cycles, the impact of its decline on global climate, and the geopolitics of the opening up of the Arctic for further oil and gas extraction are all inextricably entangled. If an essential feature of the climate crisis is its inherent transdisciplinarity, then, surely, carefully curated stories have a central role to play in an effective pedagogy of climate change.

Let us consider, first, the story of The Scientist and the Elder. Figure 5.1 illustrates, via a concept map, the interconnections between the sea ice, the ecosystem and the people who depend on it, and the larger climatic and economic forces. This illustrates the transdisciplinarity of the story, as we can

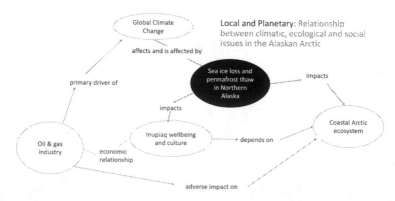

Figure 5.1 A concept map for the story The Scientist and the Elder.

recognize several disciplines even in this skeletal concept map. However, the story allows us to traverse not just disciplines but also scales of space and time. Implicit in this diagram is the history of settler colonialism experienced by the Iñupiat, as well as their complex relationship with the oil industry. To understand the melting of the Arctic, we need to go back in time, through both satellite observations and ancestral memory, to ask whether such a thing has happened before and why it is happening now.

Crucially, this story foregrounds the local – the experiences of climatic and socio-economic changes on the ground, located in place, which is where we viscerally experience weather and climate. The local is also the scale at which action and decisions can be taken for adaptation and mitigation. Centering the local is counter to the mainstream discourse on climate change, which is inevitably top-down, and which, through its global, quantitative averages, tends to wash out the felt, on-the-ground, place-specific predicaments of people and non-humans. Consider one key feature of the climate problem at the local scale: the high degree of uncertainty in projected changes in temperature, precipitation, and weather patterns. As computer models of climate downscale from regional and global scales to the local, uncertainties become larger and larger. This results from the complex nature of the climate system, entangled with uncertainties in socio-economic drivers of change. In a volume about the politics of climate change and uncertainty in India, the editors argue that 'theorising about climate-related uncertainty by experts, modellers and policymakers may have very little to do with how local people (men, women, third gender who are, in turn, differentiated by age, ableism, class, caste, location and ethnicity) make sense of climate change and live with climate-related uncertainties in everyday settings …' (Mehta, Adam, and Srivastava 2021). Further, even when multiple kinds of knowledge are admitted by those who have privilege and power, local knowledge is often ignored, sidelined, or appropriated by dominant knowledge systems (Movik, Synne et al. 2022).

> A certain politics of knowledge and valuation results in particular domains (especially so-called hard science) gaining authority over the others. Yet, all forms of knowledge (including so-called expert knowledge) are culturally and socially embedded and moulded by particular social, power and gender relations. Models are also embedded in narratives and storylines about a future based on certain assumptions … but through a range of political practices and boundary-ordering devices gain authority over other forms of knowledge …

Therefore, while both local and global scales are important, starting with the local and tracing connections between local and global and back again prevents us from privileging the mainstream top-down discourse. That this is a matter of justice and power we'll see in more detail for the example of the Alaskan Arctic in Chapter 6, where we explore the larger context of the story of The Scientist and the Elder. Through this exploration, students are able to get a sense of the density of interrelationships hinted at in the concept map: relationships between local and global, human and non-human, economic imperatives and ecosystem changes.

Justice issues manifest in multiple ways as we explore the context of this story. First, the Iñupiat have contributed little to the problem of climate change, yet they suffer disproportionately from its impacts. As we'll see in the next chapter, the cultural significance of sea ice, its role in survival and well-being, and the corresponding risks due to its rapid melt make clear this aspect of justice. A short exploration into the Alaska Native Tribes' struggle for land rights under settler colonialism helps clarify the historical context as well as the present-day complex relationship of the Iñupiat with the oil industry that is its main source of revenue. A wider examination of Indigenous knowledge systems leads to an understanding of epistemic injustice: that colonialism and marginalization not only inflict suffering in terms of dispossession, quality of life, and livelihoods, but also crush alternative paradigms and onto-epistemologies, other ways of thinking and being – place-based, local knowledges. Additionally, the plight of other species, which suffer the consequences of Arctic warming, allows us to consider multispecies justice.

The significance of the story of the Mouse on the Tundra becomes clear when we think about dominant paradigms. It is, of course, a traditional teaching tale about how apparent limits and pre-conceptions prevent us from seeing the world in its fullness. It is a story not about mice but about human hubris and foolishness. Having explored the Newtonian paradigm in the physics context and widened our understanding to include its influence on Modern Industrial Civilization, we can now interpret the Mouse and the Tundra story as a cautionary tale about a particular kind of non-seeing that I call *paradigm blindness*. The dominant paradigm is the air we breathe; its unquestioned assumptions shape our sense of what's 'normal,' our ideas of the good life, our customs and assumptions about economics, about our relationship with the rest of Nature. Because these constitute our default reality, we are not generally aware that the dominant paradigm is a *construct*, and, therefore, it can be deconstructed, shifted, transformed. This discussion is, of course, situated within the epistemological dimension of this pedagogical framework. Thus, three of the four dimensions are made apparent through the story and its wider context.

Let us now examine the story of The Village Women and the Forest (Figure 5.2). At first sight, this story has nothing to do with climate change. Indeed, the main threat to these villagers is not the changing climate, but deforestation and desertification of the landscape, resulting in lowered water tables, drying streams and rivers, dwindling forest resources, vanishing species, increased incidence of malaria, and hotter summers. Like vast numbers of people in India and the world, these villagers live off subsistence agriculture and forest produce, relying on the forests also for medicinal plants. Water security is paramount for their survival and well-being.

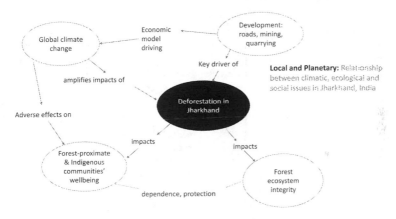

Figure 5.2 A concept map for the story The Village Women and the Forest.

Climate change enters when we consider the wider context of this story. The state of Jharkhand, where this village is located, is among the most climate-vulnerable regions in India. This is because climate projections for the region include heat waves, increased rainfall variability (always bad news for rain fed agriculture), diseases, and increase in human-animal conflict – and also because Jharkhand ranks lowest among Indian states in terms of the Human Development Index. As climatic changes manifest in the region, it is the deforested, desertified areas that will see the worst impacts. It is against this backdrop that we must view the story of Parvati Devi and her community of forest restorers.

The story of The Village Women and the Forest that I told in Chapter 1 reminds us that climate change is not the only problem that we are up against. In fact, climate change is not the foremost issue for many marginalized communities – they often have far more immediate and dire concerns to deal with, from hunger and poverty to dispossession. A 2019 interview series conducted by activists to inquire into how marginalized communities in India thought about climate change revealed that forest-dwelling or forest-proximate communities in India were more concerned about the immediate threat of dispossession resulting from a legal challenge to a law that protected their traditional rights to forest resources (People's Climate Network 2020). Some communities, such as fisherfolk, who are directly affected by climate impacts such as sea level rise and changes in ocean currents, and are keen observers of these changes, tend to articulate concerns about climate change more than others (Yuvan Aves, private communication, February 9, 2023).

So, what are the Village Women in the story up against? Belonging to an impoverished community of tribal and lower-caste peoples, they face social discrimination and marginalization, and are also subject to neglect or violence from the state. More than two decades ago, when large-scale razing of forests commenced in their area, they noted the disappearance of several species, including tigers, the change in local weather, with summers getting much hotter, the fall in the water table, the drying up of streams, and a rise in the incidence of malaria. As the region became desertified, there was more pressure on the last remaining forest, about 200 hectares, which began to dwindle. It was then that Parvati Devi and her fellow village women decided to take action. Patrolling the forest, self-limiting use of its resources, making mud and stone check dams to conserve water, digging ponds for wild animals so that they, too, could have water to drink – all these efforts eventually paid off. When I spoke with Parvati Devi on the phone in 2019 (thanks to a friend who was visiting the region), she told me that the work continues, that she and the other village women still patrol the forest more than 20 years later. The forest has regenerated – tree trunks have thickened, many species have come back (but not the tigers), temperatures are more bearable, the malaria threat has receded, and, crucially, the water table has gone up sufficiently so that agriculture is possible again. Parvati Devi spoke with passion and ebullience. She was lucky, she said, that she only had to travel one kilometer to the village pump for her daily water needs. Typically, women in village India where the local ecology has been ruined walk several kilometers daily to collect water. 'Come and visit us,' she said to me. 'Come, see what we have done!'

A physics course leaves little room for a deep dive into the historical context of the story of The Village Women and the Forest. However, even a limited exploration of the context, aided by the concept map, makes a few important things apparent. With a little elucidation of the history of British colonialism in India, the now globalized imposition of an exploitative, extractive, and highly unequal model of development, and the historic and ongoing suppression of the rights of tribal and other forest-proximate peoples, along with gender oppression, the actions of Parvati and her fellow forest protectors stand out as both brave and poignant. The wider context of Jharkhand's climate vulnerability makes it clear that community protection of forests is crucial to resilience from climate impacts. The once nearly contiguous great forests of Jharkhand are now fragmented islands in a growing desert while the dust of stone quarries extends kilometers over the terrain. Fortunately, Parvati's story is not unique. In Jharkhand and other parts of India and the world, communities are protecting forests and other natural habitats with which they have developed deep relationships over centuries or millennia. While people like Parvati have done the least to contribute to the climate problem, they are likely to be impacted the worst by its impacts. This is centrally what climate justice is all about, but it goes deeper. Parvati and other protectors are rarely acknowledged by the rest of the world. Their experiences do not inform policy, nor do these women get a seat at the table in international negotiations. Their creativity and courage against great odds are rarely celebrated in glossy magazines

or through international environmental awards. Nor are their ways of conceptualizing the world given serious consideration. Against the Dominant Social
Paradigm of neoliberal societies (Bogert et al. 2022), their knowledge systems and ways of living and being are dismissed as primitive or unsophis-
ticated, if they are acknowledged at all. Inequity and a colonialist ethos are
manifested also when, for example, large development agencies fund projects
for marginalized communities to adapt to the impacts of climate change –
these top-down approaches can often make people *more* vulnerable to these
impacts, a phenomenon known as maladaptation (Lisa Schipper et al. 2021).
Thus, the climate problem is also an issue of epistemic injustice with roots in
colonialism.

'What Killed Stacy Ruffin,' the third story I relate at the beginning of
Chapter 3, is a little closer to home for my American students. I divide the
article into read-aloud sections, and students take turns to do so. Afterwards
I project the photograph at the top of the story on the screen. The account of
Stacy Ruffin's tragic death, interspersed with interviews with climate scientists, foregrounds the ethical question at the heart of an extreme weather event
that is made more likely by climate change: who or what is responsible for
her death? Is it poor city planning and infrastructure? Fossil fuel executives?
All those who drive cars and use electricity? The story challenges stereotypes
about African Americans, and also opens space for a discussion on systemic
racism and environmental racism. It is also intensely moving; reading it aloud,
the entanglement between one woman's tragic death by drowning and the
socio-economic system in which we are embedded becomes viscerally clear.

Therefore, all three stories, while spanning three of the four dimensions of
this pedagogical framework (scientific-technological, transdisciplinary, and
epistemological) also illustrate how justice and power are at the heart of the
climate crisis. The sketch below by artist Michele Gutlove illustrates how
marginalized peoples and nonhuman lifeforms form the base of a pyramid of
power where wealth and resources flow inexorably to the top.

Having understood how justice and power play out at local scales through
these stories, we can now pan out to understand its global manifestations. Our
modern, globalized, extractive socio-economic system works for the dubious
benefit of a few at the expense of vast numbers of 'dispensable' humans,
non-human species, ecosystems, and landscapes. The multiple large-scale
manifestations of injustice and power include:

1   Western nations' historical responsibility for carbon emissions that created
the climate problem; note also that the Industrial Revolution was fueled
not just by coal and oil but also by resources from colonized countries that
were appropriated by violence (Hickel et al. 2022). Developing countries
have demanded from Western nations funds for these countries to develop
green energy infrastructure and to compensate for loss and damage due
to severe climatic impacts. At this moment of writing, Western nations,

including the United States, have finally agreed to a loss and damage fund; yet, this agreement is not legally binding and an agreement does not guarantee compliance (Plumer et al. 2022)

2 The fact that the richest 10% of the world's humans are responsible for 50% of individual-based carbon emissions (Gore 2020); this does not include the carbon emissions from the socio-economic infrastructures that they build and maintain

3 Indigenous people face some of the worst impacts of climate change as well as oppression and dispossession, yet their ways of knowing are also crucial as we confront climate disaster (Etchart 2017; McGregor, Whitaker, and Sritharan 2020). Indigenous groups have been active in climate action but have not been given their due place at the negotiating tables, furthering colonialist, epistemic injustice

4 People of color are disproportionately impacted by climate change, not only in the Global South but also in Western countries. As climate justice activist Elizabeth Yeampierre says (Beth Gardiner 2020), 'Climate change is the result of a legacy of extraction, of colonialism, of slavery'

5 Young people have not lived long enough to contribute significantly to the climate problem – yet, they are inheriting a world that is spiraling toward catastrophe, and they will be impacted by it the most; for the existential threat to their future, climate change is therefore also a matter of intergenerational justice. The growing youth climate movement and youth action on law and policy testify to this (Kotzé and Knappe 2023)

6 Research indicates that women and girls are disproportionately affected by the health impacts and extreme weather caused by climate change (Dunne 2020); therefore, climate change is an issue of gender injustice

7 Only about 100 corporations, including public sector companies, are responsible for 71% of carbon emissions (Griffin 2017). Some have been implicated in covering up their own scientists' research on the consequences of their actions, as well as confusing the public and stopping or delaying meaningful climate action (Oreskes 2011)

8 Through an approach called Weather Attribution, climate scientists are now able to determine the climate footprint in an extreme weather event, in the sense of a probabilistic assessment of the likelihood of that event due to climate change (see the story What Killed Stacy Ruffin in Chapter 3). While not all extreme weather events are related to climate change, a 2022 analysis of multiple extreme weather attribution studies (Carbon Brief 2022) has some grim statistics. Seventy-one percent of the 504 extreme weather events and trends in the analysis were found to be made more likely or more severe by human-caused climate change. Many (not all) of these extreme weather events occur in the Global South, in nations that don't have the infrastructure to deal effectively with them; in some cases, the event is so catastrophic that adaptation is impossible, as in the Pakistan floods of 2022

9 Multiple planetary boundary violations, including climate change, land system use, and loss of biosphere integrity, which have launched the sixth great mass extinction on our planet, underline the fact that climate change is also a matter of multispecies justice. The mass suffering and extinction (at rates far above normal) of multiple nonhuman beings are generally in the background of climate concerns, but can widen the lens with which we view the catastrophe and humanity's problematic relationship with the rest of Nature (Tschakert et al. 2021)

It is important to note that discussing categories of the marginalized in a dis-connected way is not helpful. To understand the multiple, complex ripple effects and interactions between ecological harm and human social justice issues, an intersectional lens is crucial (Yale Sustainability 2022). Stories such as The Village Women and the Forest are especially helpful in this regard, as we can see how the actions of these women resist power structures based on multiple hierarchies of gender, class, and caste. There is a lot more to environ-mental justice than what I have elucidated above (Menton et al. 2020).

As we discuss the local and global contexts, a question emerges. If human-kind is confronted with so many simultaneous, interacting, related problems – what has been called a polycrisis – then, surely, there must be a root cause. This allows us to discuss the power pyramid in Figure 5.3 in more detail. Particularly for a physics class, interrogating ideas beloved of mainstream

Figure 5.3 Inequality.

Credit: Michele Gutlove.

economics, such as extractivism and endless growth, becomes a fascinating exercise. As I've pointed out in a book chapter co-authored with climate physicist Chirag Dhara (Dhara and Singh 2021), endless economic growth defined in terms of GDP ignores limits and constraints set by the laws of physics and the mathematics of exponential change. For example, while recycling is absolutely necessary, there are physical limits to what can be recycled and how much – similarly, there are hard limits to efficiencies of technologies. While we urgently need to move away from fossil fuels to alternative energy, simply replacing gasoline-powered vehicles with electric vehicles will not solve our polycrisis, for which a system change is necessary. When we consider not just the carbon footprint but also the material footprint – the sum total of all material, from extraction through production, use and disposal, that is used to produce any technology – and learn that the global material footprint continues to rise dangerously every year (United Nations n.d.) it becomes possible for students to develop a nuanced, critical perspective on technofixes and the necessity of system change. (I elaborate on this in Chapter 7, where we discuss real and false climate solutions). Apart from the injustice and harm to multiple humans and nonhumans, our globalized economic system is simply not tenable. Economists such as Kate Raworth on doughnut economics, which posits both an environmental ceiling, based on planetary boundaries, and a social foundation for human well-being, and Dan O'Neill, who posits an Economics of Enough, provide two examples (Raworth 2018; O'Neill *The Economics of Enough* 2014) of a range of attempts to critically re-examine economics and consider alternatives to the current system. The three stories I have mentioned enable us to explore economic injustice and the exploitative nature of the current system, and the reasons why alternatives are needed.

5.2   Stories and the Three Meta-Concepts

So far, I have attempted to show that carefully curated stories reflect the transdisciplinary nature of the climate crisis as well as the centrality of power and justice. These are two essential features of the climate problem, as I elaborated in Chapter 1. The other two key features: large scales of space and time, and inherent complexity, also become evident in our explorations. Paleoclimate and human histories matter when we ask how we got to our fraught present. Connections across space become clear when we ask if the stories are connected to each other, and possibly to us in our current locales. The storm system that lead to extreme flooding and the death of Stacy Ruffin had its origins in regional weather phenomena that are connected spatially across the globe. Further, the climatic footprint in this storm relates it to the Industrial revolution, colonialism, and the birth of Modern Industrial Civilization. We can also ask: does melting sea ice in the Arctic have anything to do with heat waves in Jharkhand? In the next chapter, we will discuss how Arctic melt has global consequences, not just exacerbating global heating but also affecting regional phenomena. Recent research indicates possible links to increased risk of wildfires in the Western US and the drying out of the Amazon rainforest. These climatic teleconnections indicate a world of deeply related, dynamically interacting complex systems.

Complex interconnections are, of course, one of the three meta-concepts that provide a conceptual structure for making sense of climate change. The other two, Balance/Imbalance and Critical Thresholds, are also apparent in these stories, certainly when we pan out to the global scale to understand the climate problem, as Figures 5.1 and 5.2 indicate. In the story The Village Women and the Forest, it is also possible to reflect on local shifts from relatively steady states to change through the crossing of local thresholds, ecological and sociological. Local ecological shifts represent the transition from an average Balance to Imbalance (with the cautions regarding the use of these terms described in Chapter 4). We know that a consequence of deforestation is fragmentation – the marooning of small islands of forest within human-modified landscapes. It is well known that fragmentation speeds biodiversity loss and affects nutrient cycles (Haddad et al. 2015). The lowering of the water table due to deforestation leads to loss of agricultural productivity, and eventually destabilizes the community. When a resilience threshold is crossed, distress migrations result, with young men leaving to join the urban poor in distant towns and states. It is harder to find *local* aspects of the meta-concepts in the media story of Stacy Ruffin without exploring the wider context, but all of these stories are about change – long-term background changes as well as sudden shifts in local conditions, both societal and environmental. Critical thresholds manifest as transition regions that commit systems to change in one direction. How we encounter and deal with change has to do with resilience, community, knowledge systems, history, governance, and our position in the pyramid of power. I explore in detail how these meta-concepts come into play in the context of the Arctic in Chapter 6. At the global level, critical thresholds include so-far-identified nine planetary or Earth system boundaries. These stories, especially The Village Women and the Forest, expose the ways in which climate change interacts with other violations of planetary boundaries, such as species extinction and land use change.

In the classroom, these stories and others like them engage the whole student: not just the intellectual but also the emotional self. Students remember these stories each time I bring them up, even if weeks have passed. I also have a sense of student engagement on the basis of their responses in discussions and assignments. Transformative education theory indicates that true learning results in an epistemic shift – a cognitive-affective change in the way the student constructs the world such that the student cannot go back to being the same person as before. I have no way to measure such a long-term shift within the period of one semester. It is possible that an epistemic shift contains not one, but many thresholds over a long period. I have certainly witnessed perspective shifts in students within the duration of the semester, but I cannot say how long they last, except on the rare occasions when I happen to hear from former students. It is likely that some impacts of my teaching approach are lost over time, or are diluted by other, more conventional classes, which is one of several reasons why we need a systemic change in education – at the very least, some kind of collaboration across disciplines and courses.

5.3   Stories as Ontological Tools

I will demonstrate, based on my elucidation above, that my choice of stories constitutes what researchers in Science and Technology studies refer to as boundary objects (Star and Griesemer 1989) I quote:

> Boundary objects are objects which are both plastic enough to adapt to local needs and the constraints of the several parties employing them, yet robust enough to maintain a common identity across sites. They are weakly structured in common use, and become strongly structured in individual-site use. These objects may be abstract or concrete. They have different meanings in different social worlds but their structure is common enough to more than one world to make them recognizable, a means of translation. The creation and management of boundary objects is a key process in developing and maintaining coherence across intersecting social worlds.

Stories like the ones I have chosen are portal pathways across disciplines, as they are at the nexus of the natural and social sciences. We can think of them 'in common use' as simply stories, but viewed from different disciplinary perspectives, they acquire specific, sharper meanings. In this sense, these stories are clearly boundary objects, but they also enable travel across other boundaries: local and global, space and time, human and non-human. As I will demonstrate, carefully curated stories also enable us to traverse onto-epistemological boundaries, thus serving epistemic justice.

In their seminal work, Storylistening: Narrative Evidence and Public Reasoning, Dillon and Craig (2022) argue for the power of stories in the public sphere. 'Contemporary circumstances have exposed the limits of the scope and power of scientific evidence, and the risks of ignoring or denying the power of stories to influence public opinion.' They argue for narrative literacy, which involves understanding the function and effects of stories at the individual and collective levels. According to them, the belief that stories are important for generating empathy is ill-founded; rather, stories are crucial for presenting multiple points of view. Stories also serve as narrative models for explanation and understanding. Perhaps most importantly, stories help create and consolidate collective identities, including social norms and values, and indicate how these change over time. Thus, stories are also ontological tools. Considering the discussion on paradigms and onto-epistemologies in Chapter 4, we can see that the dominant social paradigm of modern industrial civilization is shored up by *dominant narratives*: stories, implicit and explicit, that promote and justify the status quo. Governments, corporations, and advertising agencies have long employed the power of narrative for their particular ends. The challenge is that these narratives are part of the air we breathe, and so, for the most part, they go unquestioned.

I can offer two examples of how dominant narratives – reflected in popular media and also implicit in invisible stories – have shown up in my classroom. One semester, I took my class to a showing of the movie *Interstellar*. The movie is set on a dying Earth, where humans live bleak, hard-scrabble lives in the wave of a global blight that has caused large-scale dust bowl conditions. The hero, a former NASA pilot, declares nostalgically that humans used to be explorers, not caretakers, and that 'Earth has turned against us.' The idea is that humanity's future is in space, and now that the Earth has been sufficiently used up and rendered useless, this white-man-hero will lead humanity to its true destiny, to the stars. Student reaction to the film was favorable, until after our discussion about the problematic issues in the film: the colonialist/racist subtext, the relinquishing of responsibility for the plight of the planet, the escapism of the privileged. Following that, the students' perspectives acquired far more nuance. Another example relates to Elon Musk's famous gesture of throwing his Tesla into space in 2018. Initial student reaction was uncritically admiring of the techno-billionaire's actions, reflecting the pervasiveness of invisible narratives about the super-rich in America. Students were initially unable to relate this extravagant gesture to their own hard lives and the fact that some of them had to work hard to make payments on their cars, if they had cars. A discussion about the role of the super-rich in worsening climate change, their apparently limitless consumption of resources, and the growing problem of space debris around Earth helped students think more critically about Musk's behavior.

The current exploitative economic system – neoliberal capitalism – cannot be separated from colonialism, which has enabled it to become a globalized destructive force wreaking havoc on the planet. It is not just that the origins of capitalism, including the rise of the fossil fuel era, are intricately linked to colonialism (Malm 2013) but that the impact of colonialism persists to this day. Peruvian sociologist Anibal Quijano's notions of coloniality of power and coloniality of knowledge help us understand that the legacies of colonialism, in the form of infrastructure, governance, and intellectual concepts – persist long after the colonizers have left. For example, the timber-focused Forest Department, much of the penal code, and the elite model of conservation based on assumed Nature-Culture separation are relics of empire that continue to oppress and dispossess the marginalized in India. Thus, Bhambra and Newell (2022) remind us that our modern world is in fact a *colonial modern world*, 'based on the historic destruction of worlds of others,' and that any serious attention to climate change requires attention to the historical role of colonialism in creating socially and regionally uneven concentrations of wealth. Far from being over, colonialism has, in fact, shape-shifted from the nation-over-nation kind to colonialism by transnational corporation. Indigenous (Potawatomi) scholar Kyle Whyte also points to the problematic portrayals of 'Indigenous vulnerability to climate change without reference to the larger struggles with colonialism and capitalism …' (Kyle Whyte 2019). Among the things colonialism destroys is the possibility of alternative ways to live and be.

Therefore, stories from the margins can help shift the paradigm by questioning, resisting, and positing alternatives to the status quo, including alternative onto-epistemologies. Indigenous scholar Doreen E. Martinez (Martinez 2021), offering a conceptual framework that helps distinguish key elements of Indigenous storytelling, says that 'Indigenous epistemologies are practiced through Indigenous storying.' Further, such storying has disruptive power:

> Typical Western expectations reinforce individualism and a type of ownership or proprietorship, in which the receiver has ultimate power. Additionally, it is these Western/European/First World ideologies that we must also attend and disrupt.

In Chapter 6, I examine how Indigenous ways of knowing in the Arctic and beyond present challenges and alternatives to the Newtonian paradigm, and attempt to elucidate the onto-epistemological role of stories in Indigenous traditions. Meanwhile, a comparison between the values that motivate women like Parvati in forest-proximate communities in India and the value system of modern industrial civilization helps to drive home the point. The village women are motivated to take action for the sake of both people and animals, for both long-term benefits and immediate needs. In a social system with gender and caste oppressions, they are always conscious of their low social status, and yet they stand up to power in order to protect their forest. They have a sense of complex interconnections in human-nonhuman worlds through their own relationship with the forest and an appreciation for biodiversity as a necessary condition for a healthy forest. Their actions result in restoration and healing for both human communities and nonhumans. Contrast this with the (admittedly grossly oversimplified) value system of modern industrial civilization: short-term thinking, self-centeredness, exclusively human-centric, general obliviousness to power structures, and uncritical admiration for those on top; simple, linear thinking, and actions that tend to destroy rather than restore or regenerate. Contrasting value systems indicate different onto-epistemologies.

The Mouse in the Tundra story with which I began this chapter is especially helpful for students to understand paradigm blindness and its relationship to climate change. After they have discussed the other stories, been introduced to complex systems and the limitations of Newtonian thinking within and beyond physics, and understood the terms 'paradigm' and 'paradigm shift,' the meaning and significance of the Mouse's story become apparent.

5.4 Stories as Diffractive Tools

A thin, blood-red laser beam cuts through the darkness of the lab. It shoots through a tiny hole in an opaque rectangular plate and emerges on the other side, where it eventually hits a white screen, forming an image.

My students are expecting that the image on the screen will be no more than a tiny, bright red blob, the size and shape of the tiny slit through which the laser beam is passing. But what's surprising is that the image is larger than the almost imperceptibly small hole, and further, there are fringes: for a circular slit, concentric bands of light and dark on either side of the central bright spot.

A laser beam is a special kind of light beam; it is to light what a band of marching soldiers is to a crowd moving down a street toward a playing field for a game. It is useful in this experiment because it reveals a property associated with all waves: diffraction, the bending or spreading of waves around an obstacle or through an aperture. Light is a wave. The image on the screen: the central bright blob and the surrounding fringes is called a diffraction pattern (Figure 5.4).

A wave – the word in physics implies the entire disturbance, that is, the full series of crests and troughs – has the peculiar property that when it is superimposed with another wave, the two move through each other, superimpose, and 'interfere' to create a new pattern. (Contrast this with the behavior of two particles, which, unlike a wave, are localized; when they collide, they may stick together, bounce off each other, or break into pieces). This property of waves can be easily demonstrated in the kitchen sink. Fill your kitchen sink

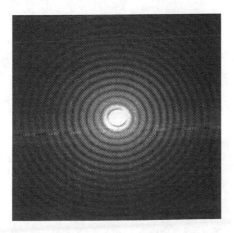

Figure 5.4 A diffraction pattern from a laser beam passing through a circular aperture.
Credit: Wisky, Wikipedia.

with water. Jab the surface of the water rhythmically by moving a spoon up
and down. A wave (a rippling series of crests and troughs) will form, widening out from where the spoon hits the water. Now, add another wave by
using a second spoon. When the two waves meet, they cancel each other out
in some places and enhance each other in other places, forming a shifting,
shivering pattern. As with water waves, light does that too, whether only one
wave is present or more than one. (One wave can become a source of multiple waves when a barrier is present.) One of the astonishing things about a
diffraction (or interference) pattern is that the dark fringes, as much as the
bright ones, are also produced by light waves! Darkness results from two light
waves canceling each other! And this is just the classical context; single-slit
and double-slit experiments with light and with tiny particles such as electrons open the doorway to the utterly strange and counter-intuitive world of
quantum physics.

But let us confine ourselves simply to the classical domain and the diffraction of light, and why the metaphor of diffraction is relevant to our
concerns. A diffraction pattern is fundamentally the result of an intra-action.
(In Chapter 10, I follow scholars in contrasting diffraction with the far more
passive optical phenomenon of reflection, which also serves as a powerful and
rather ubiquitous metaphor.) For now, it suffices to note that in diffraction,
the pattern of alternating darkness and brightness (or, for water waves, flat
regions, and high crests/deep troughs) cannot be synthesized into an average
or an undifferentiated whole, not without losing key details. The differences
and similarities in the pattern at different locations matter – they are irreduc-
ible. A crucial aspect of Karen Barad's analysis (which is inspired by quantum
phenomena, outside the scope of this book) is that the apparatus of diffraction
includes not just the laser source, the diffraction slit, the screen and the optical
bench on which they are mounted, but also the experimenter – the educator in
the room, and the wide-eyed students, and the socio-cultural-material context
in which they are embedded. Our concepts then arise from the entanglement
of the material and the social-cultural, based as much on what we exclude
from consideration. Matter matters!

The point I want to make is that the phenomenon of diffraction, as originally suggested by feminist philosopher Donna Haraway and developed in
new directions by Barad, serves as a basis for a useful metaphor. A key insight
from Barad is that ontology (concerned with what exists), epistemology (concerned with how we know what we know about existence), and ethics are not
separate, but constitute an ethico-onto-epistemology. Here is Barad on page
185 of her book (Barad 2007):

> Practices of knowing and being are not isolable; they are mutually
> implicated. We don't obtain knowledge by standing outside the world;
> we know because we are of the world. We are part of the world in its
> differential becoming. The separation of epistemology from ontology

is a reverberation of a metaphysics that assumes an inherent difference between human and nonhuman, subject and object, mind and body, matter and discourse.

When we declare that something exists, and explain how we know that it exists, we are responsible for how that knowledge reverberates in the world. Ethical obligations thus emerge when we acknowledge the inherent relationality of the universe. My brief exposition of Karen Barad's philosophy of agential realism doesn't do it justice, but for the moment, the above should suffice.

What does diffraction-as-a-metaphor entail? Scholars in education and philosophy have considered this question.

> Just as Barad described diffraction as a dispersal of waves, a diffractive analysis functions to move me away from habitual normative readings that zero in on sameness toward the production of readings that disperse and disrupt thought as I plug multiple theories into data and read them through one another. A diffractive analysis is not a reduction of data using a series of concepts, much like coding would require. Rather, it takes a rhizomatic (rather than hierarchical and linear shape) form that leads in different directions and keeps analysis and knowledge production on the move.
>
> (Mazzei 2014)

Also:

> For Barad, diffraction is a useful tool highlighting the entanglement of material-discursive phenomena in the world. Diffraction is thus predicated on a relational ontology, an ongoing process in which matter and meaning are coconstituted.
>
> (Bozalek and Zembylas 2017)

Carefully selected stories such as the ones I have highlighted in this book, when read in the way I've described (see also Chapters 6 and 10), may be considered *diffractive tools* – because they enable us to 'read through each other' the disciplinary threads entangled within the story, the contrasting paradigms of modern industrial civilization and Indigenous onto-epistemologies, the human and the non-human. (This is not disconnected with the role of stories as boundary objects and ontological tools, as discussed earlier.) Science, ethics, and justice meet as equals in the theater space of story, and boundaries become alternately blurred and sharpened. None is privileged over another, and there is no attempt at a grand synthesis; instead, the differences and similarities among disciplines and onto-epistemologies can enable the emergence of new understandings and concepts. (This is not to say that grand syntheses in the form of global averages – for climate change, for example – are not useful, but when they constitute the dominant view, they can wash out local details and undermine justice.) Further, when we storify science through enactments of physical processes, we acknowledge our belonging and participation in the universe. This is also resonant with Barad's notion of matter as active in the universe – a fact long ignored by humanists and social scientists for whom the physical universe is a passive backdrop to the human drama – and unsurprising to physicists (although our appreciation of matter is constrained within a narrow context). From the inherent relationality of the universe emerges an ethics of responsibility. Thus, a diffractive approach must also make a difference. I elaborate on these matters in Chapters 6 and 10.

5.5   Other Kinds of Stories: Storifying Science, and Science Fiction

Thus far, I have examined the multiple pedagogical uses of real life and real-life-based stories, as well as one teaching story from the Iñupiaq tradition. However, there are other kinds of stories that can enrich the understanding and delight the imagination. A physicist's view of the universe reveals narratives of the non-human that – when rendered creatively – turn out to be as fascinating as those of homo sapiens. As per Dillon and Craig, if we accept that stories are causal accounts of something happening, but set aside the requirement of agency as it is normally understood, then it becomes apparent that the universe is continually telling stories that we can interpret through the lens of science. The interactions between matter and matter, the flux of energies that animate star cores and planet formation and life itself allow us to see matter as active in the universe. Science is a particular way of eavesdropping on (and participating in) what might be called the Grand Conversation. To give a personal example: an 'ordinary' walk across the campus on a snowy day. As the snowflakes fall, I look up at the clouds and think about water molecules in the upper atmosphere coalescing, acquiring greater mass, succumbing to Earth's gravity, the same force that keeps me tethered to the planet and the planet to the sun. The snowflakes accelerate at first, but as the air ruffles their intricate lace edges, they begin to fall at a steady pace. Their occasional swirling, and the way they paint the northern sides of tree trunks, makes visible the invisible currents of the winds. The water molecules in a single snowflake are likely to have come to this particular locale from widely dispersed places: the moist exhalation of an elephant in the Congo, the evapotranspiration of a kapok tree in the Amazon rainforest, the steam rising from a stew cooking in an apartment near the Alps. This local storm is part of a weather system that is communicating with and responding to larger-scale shifts in atmospheric conditions that span the continent and the world. The shafts of sunlight that break through the clouds and make the snowfield before me glitter have traveled the empty void between sun and planet for eight minutes before falling into my eyes and making me blink. Once part of the mass of a neutron or proton, this light and warmth are transmuted matter, released from the tremendous, crushing pressure at the heart of our star. A walk through the quad becomes an experiential meditation, a belonging to something greater than our individual selves, where we are characters in a vast cosmic play. We are all part of this seething ocean of stories, and yet, most of us are blind and deaf to it as we go about our lives.

It has been my effort over nearly two decades – well before I introduced climate change into my general physics classes – to bring to life the drama of the physical universe that is playing around us. I have used two techniques to do so: one, physics stories, wherein students share, biweekly, an everyday experience with a physical phenomenon that might be 'ordinary' but leads to discussion, exercises in hypothesizing, and, ultimately, the development of a keen awareness of the physical universe in which we are emplaced. My second technique is what I call 'physics theater,' in which we use embodied or kinesthetic learning to enact physical processes. In my first such experiment, motivated in part by persistent student misconceptions about electric currents, I drew a giant circuit on the classroom floor, assigned a student to act as the switch, and another to be the battery, while other students volunteered to enact electrons traveling the circuit. I found that when performed appropriately, this 'physics theater' helped clarify conceptual ideas for students and prevent misconceptions. Additionally, students enjoyed themselves and came to see otherwise-remote-from-experience entities like electrons as present and active in the world around them. These activities also helped breach the conventional subject-object separation in science toward more of a participant-observer 'intra-action' with matter (to use the neologism from physicist-philosopher Karen Barad). Embodied learning is, in fact, a subject of scholarly study, as I later discovered (Euler, Rådahl, and Gregorcic 2019). I was convinced of its power when, much later, during an NSF-funded large-team project in which I was the physics teacher and curriculum co-developer, we taught physics concepts to African-American schoolgirls through dance. The girls were talented students of dance, and created beautiful choreography around physical processes, from the formation of the solar system to the oscillations of the carbon dioxide molecule.

To storify climate change, we have to think of the nonhuman as active in the universe, possessing agency of a different kind than the way we usually understand the term (as per physicist-philosopher Karen Barad's agential realism (Barad 2007)). Consider the greenhouse effect, one of the fundamental phenomena that helps us understand the role of carbon dioxide and other greenhouse gases in global heating. While it is not difficult to comprehend, it has details that can be confusing for the novice. Mixing up the behavior of terrestrial heat waves with that of carbon dioxide molecules, assuming that only solar heat is absorbed by the Earth, and having problems remembering the process a few weeks after being introduced to it are a few of the issues that arise. However, when I introduced the greenhouse effect as a narrative to be enacted, with the help of props (appropriately colored streamers for the sun's and Earth's radiation, for example), the results were encouraging. A freshman seminar on Arctic climate change became a testing ground for the efficacy of this enactment. Students took on various roles: the sun, the sun's radiation, the Earth's different surfaces, from oceans to ice, the Earth's radiated heat waves, and carbon dioxide molecules in the atmosphere. With some help from me, they wrote the words for each character, and a central open space in the middle of the classroom became the stage. On another occasion, I taught the greenhouse effect via an enactment to a group of village children ranging from ages 12 to 16 in the Indian Himalayas, during a study abroad winter break experience with American college students. The students helped the children to understand and rehearse the enactment, and they then performed it. Their immediate, intelligent understanding of it, their enthusiasm for the enactment, and the connections they could make thereafter with their local environmental issues were encouraging, although as an isolated exercise I do not expect it would have lasting impact.

Storification in climate science – harnessing the power of narrative to make climate science meaningful for local communities, municipalities, and decision-makers – is one of the most interesting developments in recent climate research. Climate scientist Ted Shepherd and colleagues have proposed physical climate storylines: 'A physically self-consistent unfolding of past events or of plausible future events or pathways' (Shepherd and Lloyd 2021). An example cited is the destruction of Mackenzie river delta freshwater ecosystem in Canada, after a severe storm; presented as a storyline, we can see causal connections between global climatic change, such as sea ice loss and sea level rise, and local conditions. Shepherd's motivation for storylining is to address Sheila Jasanoff's critique that conventional climate science divorces knowledge from meaning through abstraction; storylining attempts to close the gap between the production of climate information and its use by communities. Because climate adaptation is inevitably local, and because climate models, while excellent for global or large-scale projections, still have large uncertainties when downscaling to the local scale, climate storylining presents a causal, sense-making narrative that reflects all the complexities of local conditions and their connection to global phenomena while also connecting the human and social with the climatic and ecological. Through causal maps and what-if explorations of alternative possibilities, storylines embrace local complexity while enabling narrative simplicity. Storylines represent an epistemological shift from conventional climate information; being 'plural, conditional,' they avoid the kind of single, definitive declaration that policy makers expect from scientists, but which are impossible when uncertainties are large (as they tend to be for very localized projections). They also help connect the local with the global, as a causal map of the changes in the Mackenzie river delta ecosystem (reproduced from the referenced paper) indicates. The possibility

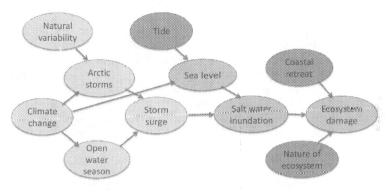

Figure 5.5 A causal network based on a physical climate storyline about the Mackenzie
      Delta river system.

of welcoming multiple voices and knowledge systems is also greater with
storylines (Figure 5.5).

Other climate narrative approaches that have emerged from physical climate science recently include climate risk narratives (Jack et al. 2020) and
climate process chains (Daron et al. 2019). Again, this epistemological broadening in climate science is motivated by a desire to empower communities
(Rodrigues and Shepherd 2022) to make useful decisions from climate information. In all these, we see an attempt to foreground the local, so that it can
inform, and talk back to the global discourse.

A final category of story that I will briefly discuss is climate fiction, specifically where it overlaps with speculative fiction. I use speculative fiction
as a general term embracing science fiction, fantasy, magic realism, etc., that
is, imaginative literature that goes beyond what we commonly understand as
reality. Speculative fiction – especially science fiction at its best – can rise
beyond colonialist, frontier narratives in space to accomplish two tasks. One,
to use extrapolation of current trends to play thought experiments with the future, often producing dystopic narratives, which can serve as cautionary tales.
Two, to ask the question: 'what if things were different?' to interrogate current
constructions of reality, dominant paradigms and ways of knowing, and *then*
to take the next step: immerse us in better possible worlds through the power
of story. (This, to me as a writer, is the radical potential of speculative fiction
(Vandana Singh 2021).) By placing us in worlds where different paradigms,
conceptions, and relationships hold, the best of these stories make the invisible scaffolding, ramparts, and gantries of the dominant paradigm visible and
contestable. We are, for a while, freed from the trap of the imagination that
prevents us from seeing what's outside the dominant paradigm. We are freed
from the Newtonian box in Figure 4.8. This can allow us to question current

power structures, and experience, for the duration of the story, what it might be like to live in a world that, while just as complex as our own, is more equal, more just, and more ecological sound. A good story can enable alternative possibilities to feel real, visceral, and not mere abstractions. In this way, speculative fiction can be a powerful ontological tool, countering the status quo and enabling us to dream and act differently from the established norms.

In the classroom, I have not used speculative fiction sufficiently to explore its power in detail, but I can recommend three stories that have been helpful. One is Isaac Asimov's 'Nightfall,' a short story that demonstrates (despite lack of literary quality) dramatically how devastating a cosmological challenge to a dominant paradigm can be for a society. When used in conjunction with our exploration of paradigm shifts in science and beyond, the concept becomes clear and serves to illuminate why people might have extreme reactions to violations of their paradigms. Another is Carrie Vaughn's 'Amaryllis,' an elegantly written short story about a radically different conception of family in an ecologically devastated world. Nnedi Okorafor's 'Spider the Artist,' drawing upon oil companies' devastation of Nigeria, tells the story of the relationship between a woman and a robot created by the oil company against a backdrop of repression and privation. There is a growing body of literature from humanities scholars on the uses of speculative fiction and climate fiction in the context of climate education (Rebecca L. Young 2022).

In Chapter 7 on the psychosocial action dimension, which includes critical and ethical thinking about climate solutions, I describe exercises I have done in and outside the classroom on speculative futurism: harnessing the power of story to imagine – and take back – the future. The future has already been colonized by the same forces that have historically wreaked havoc on our planet – we only have to consider the mind-boggling scale of greenwashing that is currently under way, as corporates and governments strive to maintain the status quo – and indeed, to co-opt the discourse – as climate change and other violations of planetary boundaries threaten our collective survival. Many purported climate solutions cannot stand scrutiny when viewed through the twin lenses of justice and efficacy. Speculative futurism exercises can help students nurture that critical and generally neglected faculty – the imagination.

References

Barad, Karen. 2007. *Meeting the Universe Halfway: Quantum Physics and the Entanglement of Matter and Meaning.* Durham, NC: Duke University Press.

Bhambra, Gurminder K., and Peter Newell. 2022. "More than a Metaphor: 'Climate Colonialism' in Perspective." *Global Social Challenges Journal* 1 (aop): 1–9. https://doi.org/10.1332/EIEM6688.

Bogert, Jeanne, Jacintha Ellers, Stephan Lewandowsky, Meena Balgopal, and Jeffrey Harvey. 2022. "Reviewing the Relationship between Neoliberal Societies and Nature: Implications of the Industrialized Dominant Social Paradigm for a Sustainable Future." *Ecology and Society* 27 (2). https://doi.org/10.5751/ES-13134-270207.

Bozalek, Vivienne, and Michalinos Zembylas. 2017. "Diffraction or Reflection? Sketching the Contours of Two Methodologies in Educational Research." *International Journal of Qualitative Studies in Education* 30 (2): 111–27. https://doi.org/10.1080/09518398.2016.1201166.

Carbon Brief. 2022. "Mapped: How Climate Change Affects Extreme Weather around the World." Carbon Brief. August 4, 2022. https://www.carbonbrief.org/mapped-how-climate-change-affects-extreme-weather-around-the-world/.

Daron, Joseph, Laura Burgin, Tamara Janes, Richard G. Jones, and Christopher Jack. 2019. "Climate Process Chains: Examples from Southern Africa." *International Journal of Climatology* 39 (12): 4784–97. https://doi.org/10.1002/joc.6106.

Dhara, Chirag, and Vandana Singh. 2021. "The Elephant in the Room: Why Transformative Education Must Address the Problem of Endless Exponential Economic Growth." In *Curriculum and Learning for Climate Action: Toward an SDG 4.7 Roadmap for Systems Change*, Eds. Radhika Iyengar and Christina Kwauk, 120–43. Leiden: Brill. https://doi.org/10.1163/9789004471818_008.

Dillon, Sarah, and Claire Craig. 2022. *Storylistening: Narrative Evidence and Public Reasoning*. Abingdon, Oxon: Routledge. https://www.routledge.com/Storylistening-Narrative-Evidence-and-Public-Reasoning/Dillon-Craig/p/book/9780367406738.

Dunne, Daisy. 2020. "Mapped: How Climate Change Disproportionately Affects Women's Health." Carbon Brief. October 29, 2020. https://www.carbonbrief.org/mapped-how-climate-change-disproportionately-affects-womens-health/.

Etchart, Linda. 2017. "The Role of Indigenous Peoples in Combating Climate Change." *Palgrave Communications* 3 (1): 1–4. https://doi.org/10.1057/palcomms.2017.85.

Euler, Elias, Elmer Rådahl, and Bor Gregorcic. 2019. "Embodiment in Physics Learning: A Social-Semiotic Look." *Physical Review Physics Education Research* 15 (1): 010134. https://doi.org/10.1103/PhysRevPhysEducRes.15.010134.

Gardiner, Beth. 2020. "Unequal Impact: The Deep Links Between Racism and Climate Change." Yale E360. June 9, 2020. https://e360.yale.edu/features/unequal-impact-the-deep-links-between-inequality-and-climate-change.

Gore, Tim. 2020. "Confronting Carbon Inequality: Putting Climate Justice at the Heart of the COVID-19 Recovery." Oxfam. https://oxfamilibrary.openrepository.com/bitstream/handle/10546/621052/mb-confronting-carbon-inequality-210920-en.pdf.

Griffin, Dr Paul. 2017. "CDP Carbon Majors Report 2017." CDP. https://cdn.cdp.net/cdp-production/cms/reports/documents/000/002/327/original/Carbon-Majors-Report-2017.pdf?1501833772.

Haddad, Nick M., Lars A. Brudvig, Jean Clobert, Kendi F. Davies, Andrew Gonzalez, Robert D. Holt, and Thomas E. Lovejoy, et al. 2015. "Habitat Fragmentation and Its Lasting Impact on Earth's Ecosystems." *Science Advances* 1 (2): e1500052. https://doi.org/10.1126/sciadv.1500052.

Hickel, Jason, Christian Dorninger, Hanspeter Wieland, and Intan Suwandi. 2022. "Imperialist Appropriation in the World Economy: Drain from the Global South through Unequal Exchange, 1990–2015." *Global Environmental Change* 73 (March): 102467. https://doi.org/10.1016/j.gloenvcha.2022.102467.

Jack, Christopher David, Richard Jones, Laura Burgin, and Joseph Daron. 2020. "Climate Risk Narratives: An Iterative Reflective Process for Co-Producing and Integrating Climate Knowledge." *Climate Risk Management* 29 (January): 100239. https://doi.org/10.1016/j.crm.2020.100239.

Kotzé, Louis J., and Henrike Knappe. 2023. "Youth Movements, Intergenerational Justice, and Climate Litigation in the Deep Time Context of the Anthropocene." *Environmental Research Communications* 5 (2): 025001. https://doi.org/10.1088/2515-7620/acaa21.

Malm, Andreas. 2013. "The Origins of Fossil Capital: From Water to Steam in the British Cotton Industry." *Historical Materialism* 21 (1): 15–68. https://doi.org/10.1163/1569206X-12341279.

Martinez, Doreen E. 2021. "Storying Traditions, Lessons and Lives: Responsible and Grounded Indigenous Storying Ethics and Methods." *Genealogy* 5 (4): 84. https://doi.org/10.3390/genealogy5040084.

Mazzei, Lisa A. 2014. "Beyond an Easy Sense: A Diffractive Analysis." *Qualitative Inquiry* 20 (6): 742–46. https://doi.org/10.1177/1077800414530257.

McGregor, Deborah, Steven Whitaker, and Mahisha Sritharan. 2020. "Indigenous Environmental Justice and Sustainability." *Current Opinion in Environmental Sustainability*, Indigenous Conceptualizations of 'Sustainability,' 43 (April): 35–40. https://doi.org/10.1016/j.cosust.2020.01.007.

Mehta, Lyla, Hans Nicolai Adam, and Shilpi Srivastava, Eds. 2021. *The Politics of Climate Change and Uncertainty in India.* 1st edition. Routledge.

Menton, Mary, Carlos Larrea, Sara Latorre, Joan Martinez-Alier, Mika Peck, Leah Temper, and Mariana Walter. 2020. "Environmental Justice and the SDGs: From Synergies to Gaps and Contradictions." *Sustainability Science* 15 (6): 1621–36. https://doi.org/10.1007/s11625-020-00789-8.

Movik, Synne, Mihir R. Bhatt, Lyla Mehta, Hans Nicolai Adam, Shilpi Srivastava, D. Parthasarathy, Espen Sjaastad, Shibaji Bose, Upasona Ghosh, and Lars Otto Naess. 2022. "Bridging Gaps in Understandings of Climate Change and Uncertainty." In *The Politics of Climate Change and Uncertainty in India.* Abingdon, Oxon: Routledge.

Oreskes, Naomi. 2011. *Merchants of Doubt : How a Handful of Scientists Obscured the Truth on Issues from Tobacco Smoke to Global Warming.* 1st U.S. edition. New York, NY: Bloomsbury Press. ©2010. https://search.library.wisc.edu/catalog/991008 9279802121.

People's Climate Network. 2020. "The First People's Climate Report." Tumblr. January 7, 2020. https://pcn.earth/post/190118087767/the-first-peoples-climate-report.

Plumer, Brad, Lisa Friedman, Max Bearak, and Jenny Gross. 2022. "In a First, Rich Countries Agree to Pay for Climate Damages in Poor Nations." *The New York Times*, November 19, 2022, sec. Climate. https://www.nytimes.com/2022/11/19/climate/un-climate-damage-cop27.html.

Raworth, Kate. 2018. "Kate Raworth: A Healthy Economy Should Be Designed to Thrive, Not Grow | TED Talk." 2018. https://www.ted.com/talks/kate_raworth_a_healthy_economy_should_be_designed_to_thrive_not_grow.

Rodrigues, Regina R, and Theodore G Shepherd. 2022. "Small Is Beautiful: Climate-Change Science as If People Mattered." *PNAS Nexus* 1 (1): pgac009. https://doi.org/10.1093/pnasnexus/pgac009.

Schipper, Lisa, Morgan Scoville-Simonds, Katharine Vincent, and Siri Eriksen. 2021. "Why Avoiding Climate Change 'Maladaptation' Is Vital." Carbon Brief. February 10, 2021. https://www.carbonbrief.org/guest-post-why-avoiding-climate-change-maladaptation-is-vital/.

Shepherd, Theodore G., and Elisabeth A. Lloyd. 2021. "Meaningful Climate Science." *Climatic Change* 169 (1): 17. https://doi.org/10.1007/s10584-021-03246-2.

Singh, Vandana. 2021. "A Speculative Manifesto." *Antariksh Yatra* (blog). October 20, 2021. https://vandanasingh.wordpress.com/2021/10/20/a-speculative-manifesto/.

Star, Susan Leigh, and James R. Griesemer. 1989. "Institutional Ecology, 'Translations' and Boundary Objects: Amateurs and Professionals in Berkeley's Museum of Vertebrate Zoology, 1907–39." *Social Studies of Science* 19 (3): 387–420. https://doi.org/10.1177/030631289019003001.

The Economics of Enough. 2014. https://www.youtube.com/watch?v=WIG33QtLRyA.

Tschakert, Petra, David Schlosberg, Danielle Celermajer, Lauren Rickards, Christine Winter, Mathias Thaler, Makere Stewart-Harawira, and Blanche Verlie. 2021. "Multispecies Justice: Climate-Just Futures with, for and beyond Humans." *WIREs Climate Change* 12 (2): e699. https://doi.org/10.1002/wcc.699.

United Nations. n.d. "SDG 12: Responsible Consumption and Production." SDG Indicators. Accessed April 8, 2023. https://unstats.un.org/sdgs/report/2019/goal-12/.

Whyte, Kyle. 2019. "Way Beyond the Lifeboat: An Indigenous Allegory of Climate Justice." In *Climate Futures Reimagining Global Climate Justice*, edited by Kum-Kum Bhavnani, John Foran, Priya A Kurian, and Debashish Munshi. New York, NY: Bloomsbury.

Yale Sustainability. 2022. "Yale Experts Explain Intersectionality and Climate Change | Yale Sustainability." July 28, 2022. https://sustainability.yale.edu/explainers/yale-experts-explain-intersectionality-and-climate-change.

Young, Rebecca L., Ed. 2022. *Climate Change Education: Reimagining the Future with Alternative Forms of Storytelling.* Washington, DC: Rowman and Littlefield. https://rowman.com/ISBN/9781666915792/Climate-Change-Education-Reimagining-the-Future-with-Alternative-Forms-of-Storytelling.

6 On Thin Ice

Applying the Framework to the Cryosphere

6.1 Introducing the Arctic: The Place, the People, and Cultural Relativism

I walked through the bright hallways of Illisagvik College, a two-year tribal college in Utqiagvik, Alaska. The walls were decorated with student artwork and Iñupiaq words and their translations. I found the office of the person I hoped to meet, but she wasn't there yet, so I decided to leave a message with the receptionist, a young Iñupiaq woman. As I spelled my first name for her, she asked me where I was from. 'I've never heard a name like yours before,' she said, smiling, bright-eyed, and curious. When I told her I was from India by way of Boston, her eyes went wide with wonder. 'I've only been south once,' she said, giving me a small shock of realization that from her vantage point within the Arctic Circle, 71.3 degrees North, practically *everywhere* on the planet was 'south.'

'Where did you go?'

'To Washington DC for a school trip.'

'Oh! How was it?'

She smiled. 'It was *weird*!'

I remember this encounter vividly because it turned my notion of what was 'normal' and familiar on its head. I grew up in Delhi when it was still a quiet, verdant capital city, its broad roads lined with ancient trees. For all its many differences of history, culture, and geography, Washington DC has more in common with Delhi than with Utqiagvik. For me, the treeless tundra, the ice roads, and the vast expanse of sea ice north of the town were beyond strange, shockingly different. But for someone raised here, with a sense of belonging nurtured by over 4,000 years of Iñupiaq history, this was home, this was familiar, this was normal.

And yet, these very different places around the world – Delhi, Boston, Washington D.C., and Utqiagvik – are intimately connected through global climate, including histories both human and nonhuman. We, who live in the 'south,' hardly ever think about the remote Arctic, and yet the Arctic is of crucial importance to the Earth's climate. This was my primary motivation for

DOI: 10.4324/9781003294443-6

designing a freshman seminar on Arctic climate change and its global implications. Such a course has the advantage of allowing me the freedom to focus
entirely on climate change with full use of transdisciplinary tools, instead of
having to integrate climate topics within course topics, as I do in my physics
classes. It can therefore illustrate features of my pedagogical framework with
more detail and clarity.

We begin with a short introduction by way of videos and photos of the
North Slope of Alaska, including the expanse of sea ice. I then tell the story
of The Scientist and the Elder (Chapter 1) and The Mouse on the Tundra
(Chapter 5). My narration is akin to a performance – I move around the room
and gesture as I speak, changing the tone, strength, and pitch of my voice.
I then ask the students what questions come into their minds.

I have mentioned these questions in earlier chapters; they are reproduced
below:

- Why did the ice break?
- How did the Elder know the ice was going to break?
- What is the meaning of the Mouse story?
- Why are we studying the Arctic? What does this have to do with us?

As students speak, I draw a concept map on the board (Figure 6.1). At this
point, the concept map consists only of the central bubble, the two stories,
and the student questions. We then discuss what we need to know in order
to explore these questions. Clearly, we need to know about the place and the

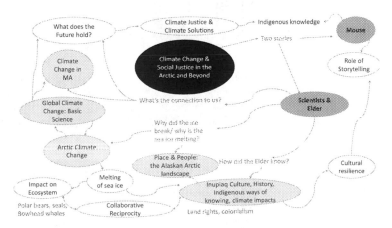

Figure 6.1 Course design according to student questions prompted by stories; the gray
areas are the bubbles introduced in the first class. The remaining concept
bubbles and links are added as we continue our explorations. Note the large
scales of space and time in this diagram.

people – the geography of the region, the culture, and history of the Iñupiat. We need to know what sea ice is and how it is being affected by climate change. An understanding of global climate is necessary to contextualize what is happening in the Arctic. And surely what's immediate and relevant to us is climate change where we are located – here in Massachusetts. Perhaps we will find some connections between the predicament of the Arctic and our own situation? At this point, I add links with question marks above them, along with concept bubbles (the two dark gray ones in the diagram) to indicate the wider context in which we must understand the stories. I tell the students that the significance of the Mouse on the Tundra story will become clearer later in the course. We return to this concept map as we explore each question, and add links and concept bubbles, so that it becomes a kind of living document of our explorations.

The Iñupiat have lived in Alaska for at least 4,000 years, engendering a relationship with the ice and the ecosystems that is nothing short of extraordinary. Modernity, settler colonialism, and capitalism have changed their lives in important ways, but certain essential aspects of their culture survive – for example, the importance of subsistence hunting, particularly whaling, which has both survival and cultural value. Looking back into the past through readings and a video (Chance, Norman 2002; AJ+ 2017) from Russia's sale of Alaska to the United States, the settling of Alaska by white outsiders in vast numbers during World War II, the outbreaks of diseases new to the tribes, the cultural trauma due to incarceration of tribal children in the notorious boarding schools run by settlers, the discovery of oil, and the long struggle for land rights – enables students to understand present-day conditions. The original relationship to the land – the Indigenous peoples of the region had no concept of private ownership – as a living, spiritual entity to which they collectively belonged, began to shift in the wake of the Alaska Native Communities Settlement Act (ANCSA) of 1971, a compromise that forced tribes to give up their claim to 90% of their original lands. Under this act, 12 Native Corporations were formed; instead of tribes' collective belonging to the land, tribal members became shareholders. Some tribes, like the Iñupiat, have benefited due to the fact that the National Petroleum Reserve of the United States overlaps their territory, and leases from oil companies provide financial benefit and jobs. Others have not been fortunate. Yet, the retreat of the sea ice, coastal erosion, and the melting tundra have not spared the Iñupiat.

When students learn about the history and culture of a people unfamiliar to them and (despite modernity) still culturally remote from them, they initially don't know what to make of customs like, for example, the seasonal bowhead whale hunts. How can the Iñupiat simultaneously consider the whale as a sacred animal, and yet hunt and eat these enormous, peaceful, intelligent creatures? Watching a video of an Iñupiat whaler describing an experience of his father's, in which the old man 'died with the whale' even as the harpoon struck – blacking out for a few moments – students' views of human-animal relationships are challenged and complicated (PBS n.d.). Here is where the danger of stereotypes about Indigenous people becomes very real: we need to avoid both the 'noble savage' stereotype and the notion that Indigenous people are somehow 'primitive' compared to us, the 'civilized.' We also need to avoid thinking about Indigenous peoples as monoliths – Indigenous cultures around the world are highly diverse, and even within communities, people do not necessarily think the same way. In this respect, I have found it immensely helpful to invite a colleague, archaeologist, and anthropologist Dr. Benjamin Alberti, to talk to my class about cultural relativism as a lens that anthropologists use to study diverse human cultures. Cultural relativism is a tool for us to understand other cultures' practices by withholding judgment and asking why, *from their own perspectives*, might people do certain things. It is an attempt to avoid ethnocentrism, which is the tendency to assume not only that one's own culture is superior but also that norms and concepts arising from our cultural paradigms are universal. Cultural relativism, Dr. Alberti points out, is not the same as ethical relativism. We can choose to disagree with certain cultural practices, but only *after* undertaking the journey to first understand them (to the extent possible) from the perspective of the other.

Animated with examples from around the world, this guest lecture, to which we refer multiple times, helps students understand that there are different ways of living and being in the world, and that the great American norm, so globalized that even my many immigrant students have internalized it, is actually only one among many possibilities. Our discussion opens the way for students who come from mixed, immigrant, or multicultural backgrounds to share their experiences of negotiating difference. Later, as we learn more about the central significance of the whale in Iñupiaq culture, the apparent paradox of hunting a sacred animal becomes at least partially resolved.

Within the first two weeks of the semester, students visit the planetarium for the experience described near the end of Chapter 3, which introduces them to the six subsystems of Earth's climate and the centrality of justice. They now have the diagram of Earth's subsystems that serves as a unifying visual tool (Figure 3.1) that I will repeatedly refer to for the rest of the semester.

6.2 Understanding Systems: Sea Ice

Having got some sense of the place and the people (I draw also from a case study I wrote for undergraduate education on Arctic climate change for the Association of American Colleges and Universities (AAC&U)), we now begin a more detailed exploration of sea ice. I begin obliquely by first introducing the concept of a *system*. I use the distinction between a heap (a collection of parts with little or no relationship between them) and a system (where parts are organized through relationships to work as a whole) (O'Connor, Joseph and McDermott, Ian 1997). Because the university experience is often bewildering for freshmen, I begin with a campus map and a tour around the campus that helps us understand what the buildings are for and how the pathways connect them. This enables students to see the university as a physical system. I then share a slide of the university's organizational structure, all the way from the president to administrators, faculty, students, and staff. Herein is the less visible aspect of the university as a system – the people and their roles, and crucially, the power differential. We discuss the ways in which some members of the university community are more powerful than others, and the presence or absence of checks and balances in the system. Our tour also includes a small community garden run by the Food and Nutrition department, and a bee habitat created by one of our biologists, in which wild bees hover over unfamiliar native plants, a stark contrast to the carefully tended flowerbeds of exotic species.

This understanding of the human and nonhuman elements of the university as a system allows me to introduce a concept map (Figure 6.2) that locates Arctic sea ice in Northern Alaska at the intersection of three systems: the ecological, the human-cultural, and the planetary. We begin with the sea ice itself.

In the late Fall, as the North pole leans away from the sun and temperatures plummet, tiny, delicate, needle-like crystals called frazil, 3–4 millimeters thick, begin to form in the frigid waters of the Arctic (NSIDC n.d.). These crystals rise up to the surface of the ocean, coalescing and thickening. As winter progresses, this ice layer floating atop the ocean can thicken to more than 30 centimeters. If it survives the summer melt, sea ice can be several meters thick, from 1 or 2 meters to even 20 meters. In winter, shorefast ice extends a few miles from the land's edge outward toward the liquid water boundary,

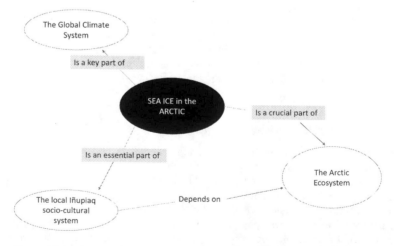

Figure 6.2 Arctic sea ice at the intersection of three systems.

always subject to the agitation of waves from below and the force of winds from above. This enables pressure ridges to form as chunks of ice pile up against each other, further thickening the ice. The photo in Chapter 1 (Figure 1.1) shows pressure ridges glittering in the April sunlight.

As the ice coalesces, pockets of concentrated salty water (brine) are created in the body of the sea ice, trapping tiny organisms: bacteria, plankton, larvae, worms. In the spring melt, these pockets connect with each other, forming intricate labyrinths called brine channels (Ask a Biologist 2014). These tiny channels (typically a millimeter in diameter) host ice algae and a number of small creatures that form the basis of the sea ice ecosystem. Later, we learn more deeply about this ecosystem, but for now, we are just starting to befriend this extraordinary entity, the sea ice.

The Northern ice cap is entirely made of frozen sea water floating atop the liquid Arctic ocean. Unlike Antarctica, which is land surrounded by water, the Northern reach of our planet is water surrounded by land. This accident of geology makes the Arctic both sensitive to, and an instigator of global climatic change, and it all begins with sea ice.

> It can be stiff and silent but also blasting and crushing with terrible noise, and it can advance and retreat as if a living being. Every year sea ice extends across 14-16 million km^2 in the Arctic (and 17 – 20 million km^2 in the Antarctic Southern Ocean) and then it shrinks and withers by the end of the summer to a fraction of its winter might. As it forms, persists, advances, and melts, it changes the ocean circulation, regulates global climate, and affects life in the polar regions.
>
> (Krupnik, Aporta, and Laidler 2010)

In the late fall and winter, as the northern hemisphere tilts away from the sun, bringing six months of near-darkness to these regions, sea ice begins to spread from the (so far) permanent ice cap over the pole, and from the shores of the surrounding lands. In late spring and summer, the sea ice melts away from the shores, and the northern ice cap shrinks. But, since satellite observations of the northern ice cap began in the 1980s, we have known that sea ice extent (Figure 6.3) and mass are diminishing at an alarming rate. As of 2023, the Arctic is warming four times faster than the planet as a whole.

A NASA video, 'Disappearing Arctic Sea ice,' explains (NASA Climate Change 2018). We see an animation of the ice (color-coded to show ice age) fluctuating through the seasons and the years, and it becomes manifestly clear that the overall extent is decreasing, and that the most dramatic decrease is in older, thicker ice. This is sea ice mass imbalance.

Essentially, what is happening is this: sea ice is created through processes like the winter freeze, and thickened by the action of ocean currents. To the east of Utqiagvik, there is an ocean current called the Beaufort Gyre that keeps sea ice circulating, a process that thickens the ice, allowing it to survive

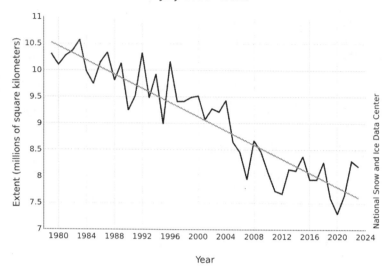

Figure 6.3 Diminishing Arctic sea ice.

Credit: National Snow and Ice Data Center.

the summer melt. Such ice, known as multi-year ice, is crucial for ensuring the build-up of future sea ice. If first-year ice is not given a chance to thicken, it may well disappear in the summer.

The processes that remove sea ice from the Arctic include summer melt as well as the floating out of sea ice through the Fram Strait on the west coast of Greenland into the warmer waters of the North Atlantic, where it melts. Thus, there are two opposing processes at work here: the processes that add sea ice to the Arctic (winter freeze and the Beaufort sea ice nursery) and processes that remove sea ice from the Arctic (summer melt and the exit of ice through the Fram Strait). When the two processes work at the same speed, the amount of sea ice in the Arctic stays constant, other than seasonal fluctuations. But today, the processes that add sea ice to the Arctic have slowed down, and the processes that remove sea ice have speeded up. As the NASA video puts it, the Beaufort gyre sea ice nursery has become a graveyard. The dramatic decline of sea ice mass and extent is therefore a local or regional example of imbalance. (I refer here to the meta-concept of Balance/Imbalance as discussed in Chapter 4.)

We now return to the concept map (Figure 6.2) that locates Arctic sea ice in Northern Alaska at the intersection of three systems: the ecological, the human-cultural, and the planetary. We are ready to explore these in detail.

6.3 Sea Ice and Global Climate

Our exploration of the link between Arctic sea ice and the global climate system (the concept bubble at the top of Figure 6.2) is a natural opening to the basic science of climate change through the three meta-concepts discussed in Chapter 4. However, because we have been studying the Alaskan Arctic region and the people who call it home, and, by now, have some sense of the local context, our subsequent exploration of global climate science is neither abstract nor at a remove. I promise students that our exploration will help answer their question as to why we are studying the Arctic. Our travel across the concept map from Arctic sea ice to the global scale will bring us back, through the meta-concepts, to the local – to the two other concept bubbles: the Arctic ecosystem and the socio-cultural context of the Iñupiat. We will also see how issues of justice and power are entwined with the biophysical.

To introduce the meta-concept of Balance/Imbalance, I follow the process I have described in detail in Chapter 4: an exercise in static balance, followed by an activity that illustrates dynamic balance. Having an embodied experience of both static and dynamic balance, we are then able to understand that balance/imbalance is about rates of change. That is, the relative speeds of the two processes determine whether we have steady state (balance) or change (imbalance). This, of course, clarifies that sea ice mass imbalance is due to large-scale imbalances in the carbon cycle and the Earth's energy budget.

As I have described in Chapter 5, one helpful technique I have used in my classrooms is embodied learning or 'physics theater' to enable a deeper understanding of physical processes, enhance enjoyment, and invite attention to the fact that matter is active in the universe. These exercises help students understand regional and global climatic phenomena as entities, actors, teachers, and storytellers. In Chapter 5, I have described students' enactment of the greenhouse effect. Common confusions and misconceptions – such as confusing carbon dioxide molecules with outgoing terrestrial heat waves ('the CO_2 gets trapped in the atmosphere') – get clarified and resolved through this process. Students' subsequent homework assignments indicate a clearer understanding of the greenhouse effect than in previous classes where I have not tried out this exercise.

The meta-concept of critical thresholds follows naturally from our discussion of global and regional balance and imbalance. What happens when dynamic balance is disturbed beyond short-term recovery? The transition region of change (relative to a chosen, relevant timescale) toward a new state represents a threshold, limit, or boundary. Our exploration of the global aspect of critical thresholds focuses on the concept of planetary or Earth system boundaries described in detail in Chapter 4. In the classroom, I teach this by first introducing the concept and the Stockholm Resilience Institute website showing the nine planetary boundaries, and explaining what they are and which ones have been crossed. I then have students study a select number (including the 6 that have been violated) in small groups, after which they present key information to the class. The two key teachings of the still-developing Planetary Boundaries concept are brought out clearly:

- There are natural limits to how much humans can exploit the Earth (often a surprise to students who are economics majors)
- Climate change is not the only serious global problem we face; there are others. Climate change exacerbates other problems – it is like an accelerant on a house already on fire. All the planetary boundaries are interconnected, so an exclusive focus on climate change (and worse, carbon reductionism) will not solve the polycrisis that we face

A key question that arises from these teachings is this: if humankind and the biosphere are beset with not one, but multiple violations of planetary boundaries, then do they have the same root cause? Why are we confronted with so many clearly inter-related problems simultaneously?

This is where considerations of power, injustice, and inequality again come into play, as I've elaborated in Chapter 5. Thus, while students first encounter these ideas in their Planetarium experience near the beginning of the course, and then again while learning about the experience of Alaska Native peoples in their fight for land rights through the painful history of settler colonialism, our exploration of global climate science through the three meta-concepts also allows us to re-visit these considerations by looking critically at the globalized socio-economic system. Although this deeper elucidation has to wait until a little later, at this point, I pose the question and ask the students to keep it in mind. The sketch 'Inequality' by Michele Gutlove serves to remind us again and again of the centrality of power and injustice underlying the multiple violations of planetary boundaries.

Let us return to the question of sea ice mass imbalance. As I have elaborated above, this is clearly related to the imbalance in the carbon cycle, which, in turn, gives rise to an imbalance in the Earth's energy budget via the greenhouse effect. Therefore, it is clear that sea ice, and the cryosphere in general, is directly impacted by global heating. But it is also the case the sea ice and the cryosphere as a whole can instigate and exacerbate global climatic changes. Our exploration of this aspect will make apparent the third meta-concept: complex interconnections.

We begin with paleohistory. The Arctic's impact on the planet throughout the last 800,000 years can be gleaned from the graph (Figure 4.4) in Chapter 4. The 100,000-year oscillations of carbon dioxide concentrations in the atmosphere represent glacial-interglacial cycles (Milankovitch cycles) where small changes in the Earth's orbit cause changes in solar radiation falling on the Arctic (although not significant change in solar radiation as a whole). This affects the sea ice mass imbalance, which, in turn, slows down ocean thermohaline circulation. The great ocean currents are a way of distributing heat from the equator (where the sun's energy falls most directly) toward the poles. When the ocean currents slow down, the equator gets hotter, causing the release of large amounts of carbon dioxide, which trigger the warm interglacial period. Currently, there is some indication that present-day sea ice melt may be implicated in the slowing down of the Atlantic Meridional Overturning Circulation, a major system of ocean currents. Thus, the Arctic is not only affected by climatic changes, but also helps instigate them.

The whiteness of ice and snow are key factors in their role in Earth's climate. Due to its shininess or albedo, sea ice can reflect as much as 80% of the sun's radiation (light, heat, and UV) back into space without absorption. (The precise percent reflected depends on the nature and surface of the ice). Reflection is a surface process in which the incoming electromagnetic wave (whether light, heat, or UV) is kicked back into space unchanged; when albedo is 100%, that much of the incoming energy is reflected back without absorption (high-quality mirrors or polished silver come close). By contrast, the darker liquid water absorbs much of the sun's energy, reflecting back only about 10%. Along with clouds, the cryosphere as a whole reflects about 30% of the total radiation falling on Earth from the sun. Thus, the sea ice, the great ice sheets of Greenland and Antarctica, and mountain glaciers all play a role in limiting how much of the sun's energy is retained in the Earth system. We can think of the cryosphere as the Earth's cooling system.

Arctic sea ice has a special role to play, however. Because it is ice afloat on water, global heating can cause the ice to melt and disappear relatively quickly. (By contrast, the ice sheets on land are extremely thick and take much longer to melt.) Consider what happens when sea ice starts to melt away due to global heating. As it melts, it exposes stretches of dark liquid water. Water will absorb, rather than reflect the sun's energy, and re-radiate it in the form of low energy heat waves. These are precisely the heat waves to which carbon dioxide molecules in the air are sensitive, and they therefore exacerbate the greenhouse effect. This leads to higher temperatures, which cause more sea ice to melt, which exposes more liquid water – and so on. This is a vicious cycle, a positive feedback loop, where the term 'positive' means 'additive,' rather than 'good.' To avoid this confusion, I use the term 'destabilizing feedback loop' for this process, which is called the ice-albedo feedback (NASA 2019). While research is ongoing to quantify the degree to which this feedback is currently active (Goosse et al. 2018), it is pedagogically useful.

It is helpful to contrast a destabilizing feedback loop with a stabilizing (or negative) feedback, such as the mammalian body's thermoregulation, or a thermostat in a house with central heating. When the house temperature falls beyond a point, the heat turns on. When the rising temperature crosses a threshold, the heater turns itself off. Thus, a roughly constant temperature is maintained. By contrast, destabilizing feedback loops like the ice-albedo feedback tend to exacerbate an initially small effect.

Unfortunately, there are many destabilizing feedback loops in the Arctic. Another important one is the permafrost-methane feedback. As the permafrost of the tundra thaws, methane gas is released, a greenhouse gas many more times powerful than carbon dioxide. Apart from creating local hazards such as sinkholes and exploding craters, methane in the atmosphere exacerbates the greenhouse effect, triggering further warming and thereby setting in motion the permafrost-methane feedback (McGee 2022). While methane is a lot less abundant than carbon dioxide in our atmosphere, and does not remain in the atmosphere for long (it eventually combines with oxygen to produce carbon dioxide and water), there are indications that the methane problem could become serious.

Feedback loops occur in both simple and complex systems, but complex systems often have multiple feedback loops that can interact with each other and potentially accelerate climatic changes. After introducing the ice-albedo feedback and the permafrost-methane feedback to students, I explain that the melting of the Arctic is making the oil and natural gas reserves on the Arctic ocean floor much more accessible. Governments and fossil fuel companies are therefore very interested in drilling for oil and gas on the Arctic seabed (and have already begun to do so in some places), which would have been very difficult earlier when the Arctic sea ice extent was greater and winters harsher. Students have little difficulty recognizing this as yet another destabilizing feedback loop. We can see, also, how the ice-albedo feedback can potentially interact with the Arctic drilling feedback loop in Figure 6.4.

Two Interacting Feedback Loops

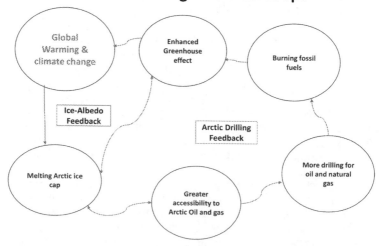

Figure 6.4 The potential interaction of the ice-albedo feedback with the Arctic drilling feedback loop.

The presence of interacting destabilizing feedback loops, along with the historical role of Arctic sea ice in the glacial-interglacial cycles elucidated earlier, helps drive home the disproportionate importance of Arctic sea ice in determining global climate. Thus, Arctic sea ice both impacts and is impacted by global climate, through a relationship of 'spiraling causality' (Tina A. Grotzer 2012). Therefore, our exploration of the meta-concept of complex interconnections in the Arctic helps answer the question: why should we care what happens in the Arctic?

Some of the central aspects of climate science that are outside the expertise of the educator can be taught through guest lectures from experts. For example, I have relied on my colleague, Dr. Amy Johnston, a geologist, for explaining ocean currents and their role in climate. Similarly, two biologists, Dr. Aviva Liebert and Dr. Brandi Van Roo, have given guest lectures on ecosystems.

6.4   Sea Ice Ecosystems

This brings us to the next concept bubble in Figure 6.2: the Arctic sea ice ecosystem. To introduce the idea of complex interconnections in ecosystems in general, we watch a full dome planetarium film called Habitat Earth, where we learn about food webs, mycorrhizal networks, hydrological cycles, and the ecological role of salmon as they travel from oceans to the streams of their birthplace to spawn. Our guest lecture from a biologist usually follows or precedes this experience. We revisit the bee habitat that was part of the campus tour early in the semester and see it again with new eyes. We contrast the rich interrelationships of the bee habitat (wild bees and native flowers, stems providing winter homes, etc.) with the conventional flower beds around campus where exotic species, unconnected with the land and its insects, are laid out, and we reconsider what we mean by 'beautiful.' We question the disproportionate importance given to individualism and independence in American society – if we depend on other species for survival, can we really say we are independent? Watching the now famous short film about the re-introduction of wolves in Yellowstone National Park, and learning about other ecosystem transitions such as the classic kelp-forest to urchin-barren transition off the Californian coast, we understand such concepts as keystone species and energy flows in ecosystems. We are now prepared for a deeper dive into understanding the coastal Arctic ecosystem, with its basis in the tiny brine channels of the sea ice.

When I arrived in Utqiagvik in 2014, upon entering the lobby of my tiny motel, I encountered a large poster, 'Beware of Polar Bears.' This warned townspeople to be wary of polar bears that sometimes venture into the streets of Utqiagvik. I have a copy of this poster which I share with my students. We pose the question: why are polar bears sometimes venturing into Arctic towns like Utqiagvik? We examine a WWF website on polar bear population and distribution around the Arctic and understand that at this point, some – but not all – polar bear populations are in trouble, earning the species the status of 'threatened.' An article from Science about polar bear metabolism and a website about Arctic ecosystems helps make sense of the connections (Ask a Biologist 2014; Whiteman 2018). Some populations of polar bears are having a difficult time finding their main food, seals. Seals make dens on sea ice to raise their pups. With sea ice thinning and retreating, denning becomes difficult or impossible. Further, seals depend on fish for food, and the fish in turn are nurtured by shrimp, copepods, and other tiny creatures nourished in turn by the ice algae that grow in the brine channels and tint the undersurface of the sea ice a yellow-green. The retreat of the sea ice means that seals can no longer rely on the sea ice for denning or the sea ice ecosystem for food.

When seal numbers drop, polar bears have to wander farther to find food. It turns out that polar bear metabolism is unusually high for a bear that size, so they use up what they eat relatively quickly. Weight loss and starvation are therefore driving some polar bears into human habitation, unusual behavior for them under normal circumstances. Thus, the metabolic imbalance in the bodies of these polar bears is directly related to sea ice mass imbalance, which in turn is a consequence of large-scale planetary imbalances in the carbon cycle and in Earth's energy budget.

This web of relationships helps students understand that causality takes multiple forms and scales in complex systems. We are used to simple linear causality – in classical physics, for example, kicking a ball – applying a net force to an object – causes it to accelerate. Cause and effect are immediately obvious. But, as Tina A. Grotzer at the Harvard Graduate School of education points out in her book *Learning causality in a Complex World: Understandings of Consequence* (Tina A. Grotzer 2012), complex systems demonstrate multiple forms of causality. The kind of causal relationship evident in the examples of kelp forest decline and weight loss among polar bears is domino causality, a chain of cause and effect. It is revelatory to students that the world around them is filled with multiple kinds of causal relations, with which their education has not, until this point, acquainted them. (Of course, this elucidation of causality makes sense if we assume time is linear; for other ideas of time and causal relationships, see Chapter 10.)

The bowhead whale, sacred to the Iñupiat, is another denizen of the high Arctic (Fisheries 2023). As much as 60 feet long, with a massive head a third of its length, this peaceful baleen whale spends its entire life in the Arctic ocean, traversing it through the year in small groups. Bowheads are able to swim under sea ice and can break it with their enormous heads in order to breathe. They eat the tiny krill, which they filter through baleen plates. They may be among the longest-lived mammals, with a life-span of over 200 years. They have complex vocalizations through which they navigate, forage, and communicate with each other as they swim. Hearing is a crucial sense for these intelligent creatures, as their complex, so-far-undeciphered sound signals indicate.

Bowheads were hunted nearly to extinction – not by the Iñupiat, but by commercial whalers from the early 1800s to the 1900s. In that time, their numbers were reduced to a mere 3,000. The US Endangered Species laws of the 1970s enabled them to recover, although they are still listed as endangered throughout their range. As their numbers climbed up, the Iñupiat estimated that the population had recovered sufficiently to allow for a return to traditional hunting, an estimate that was later confirmed by scientists (as reported in Sakakibara, 2017). They petitioned for a return to traditional whale hunts, and now there is an annual quota for Indigenous peoples of the region.

After learning about the Arctic ecosystem, polar bears, and bowhead whales, we are now ready to consider the third bubble in Figure 6.2, the socio-cultural system of the Iñupiat. Having learned about their history of struggle for land rights, the impact of the ANCSA, their relationship with the oil industry, and the range of different attitudes toward offshore oil drilling, we now employ the tool of cultural relativism to understand their relationship with the sea ice and the other living beings who depend on it, most importantly, the bowhead whale.

6.5 Sea Ice and the Iñupiat; Indigenous People and Human-Animal Relationships

In their 2009 paper, 'Respect for Grizzly Bears: An aboriginal approach for coexistence and resilience,' geographers and environmental studies scholars Douglas A. Clark and D. Scott Slocombe (Clark and Slocombe 2009) quote a participant in a focus group they organized with First Nations people in the southwest Yukon, where Indigenous peoples have co-existed with grizzly bears for a very long time. My retelling of this story, which I call The Moose Hunter and the Bear, goes like this.

One day, a hunter and his friends shot a moose and proceeded to carve up the meat so that they could transport it back to their homes. Because a moose is so large, it would take two trips to take all the meat away. They noticed a bear watching them. Knowing that the bear might take over the kill while the hunters took away the first load, the hunter asked the bear not to take the rest of the meat. After loading their vehicle, they put a tarp over the remaining meat and left the head of the moose for the bear on top. When they returned, the bear had taken away the head, but had left the rest. In gratitude, they left a large piece of meat for that bear on a rock. The hunter says that a similar incident, where the men did not talk to the bear, resulted in the bear taking over the moose kill.

'Respect for Grizzly Bears' is one of two scholarly papers that my freshman students read for our exploration of human-animal relationships among Indigenous people. Clark and Slocombe are motivated to research the idea of 'respect' for other species among Indigenous people because they don't want to make unwarranted assumptions about the meaning of the term 'respect' without asking the people themselves what they mean by it. Their context is wildlife co-management by Aboriginal and non-Aboriginal people, which requires a certain degree of cross-cultural understanding. Through the words of their interviewees, we learn quite a bit about the way values and attitudes differ between mainstream culture and North American Indigenous societies. We learn from an Elder that when he encounters a bear, he talks to the animal; that humans don't invite conflict and only shoot if the bear charges them. Other practices of respect include never directly using the word 'bear' in any language and never using a term that implies human superiority to bears. Instead, honorifics that are kinship terms, such as 'Big Grandpa' are used when speaking of bears. In one region in southwest Yukon, First Nation people and bears have come to a kind of cooperative time-sharing agreement during salmon runs, whereby humans take salmon in the mornings, and bears in the afternoon.

> Implicit in the holism of Aboriginal cosmologies is the positioning of people on an equal level with all other forms of creation, including bears. From such a perspective, there is no difference between bears and people: bears and other animals are nonhuman persons. In the distant past, living beings switched between animal and human forms; therefore, humans' interactions with animals are essentially social interactions …

And again:

> In Aboriginal worldviews, reciprocity is at the heart of maintaining appropriate relationships with elements of the natural world, including both animals and humans. Appropriate human-animal relationships are based on reciprocity: animals give themselves to hunters, who must reciprocate with appropriate behaviours or else they risk offending that animal, which will not offer itself again …

Clark and Slocombe discuss the role of stories in transmitting important ecological and behavioral knowledge about bears. Broadly, there are two classes of stories: traditional tales, which are teaching tales about values, and real-life tales that are about actual encounters with bears. Additionally, oral traditions can easily incorporate new information, and therefore 'they are adaptive in the face of change.'

Students enjoy reading and discussing extracts from this paper, as it offers a glimpse of a worldview that is strikingly different from the mainstream. This experience is re-emphasized when we turn to the second reading: 'People of the Whales: Climate Change and Cultural Resilience Among Iñupiat of Arctic Alaska' by anthropologist Chie Sakakibara (Sakakibara 2017). Here, too, I focus on specific extracts, but we read these aloud in class, and I give students a scaffolding upon which to arrange their understanding.

Sakakibara says that 'Many Iñupiat identify themselves as People of the Whales because the whale has sustained them body and soul for many generations.' Despite modernity, for most Iñupiat, the whale remains a symbol of cultural continuity as well as a practical necessity for survival and well-being. The hunt, the community sharing of meat and other body parts, and the ceremonies associated with this event define what Sakakibara calls the Iñupiat whaling cycle. In that sense, she says, 'the bowhead whale integrates all elements of Arctic life: the sea, the land, the animals and the humans.'

The climatic vulnerability of the region and the uncertainty of the future have only strengthened the desire of the Iñupiat to preserve and enhance their relationship with the whale through old and new practices. Two major concepts emerge from this paper that are useful for our purposes. One is the idea of cultural resilience. Resilience is the ability of an individual or group to 'maintain and renew itself when its existence or viability is challenged or threatened.' Because freshmen – and young people in general – often have difficult challenges in life, including, for example, responsibilities for parents or siblings, working long hours in tedious jobs to pay for college, and struggling with anxiety and other mental health issues, the idea of resilience is of immediate personal relevance, and leads to interesting discussions. Cultural resilience is more of a challenge to elucidate, but it becomes apparent that social relations, relationships to the land and animals, and connections to a shared past are important in this respect.

To understand resilience more deeply in this context, Sakakibara introduces the idea of collaborative reciprocity as it is reflected in the traditional whaling practices of the Iñupiat. 'Central to the Iñupiaq notion of collaborative reciprocity is the belief that humans and animals physically and spiritually constitute each other, that the souls, thoughts and behavior of animals and people interpenetrate in a collaboration of life … For example, many Indigenous groups in the Arctic understand that animals willingly give themselves to hunters in response to receiving respectful treatment as "non-human persons."'

This, of course, echoes the learnings from the Clark and Slocombe paper, but adds a very important element. When first introduced to the idea of reciprocity, students often interpret it as 'I do something for you, you do something for me,' a rather bleak, utilitarian interpretation that is perhaps not surprising in a capitalist, individualistic culture. I have therefore found it necessary to emphasize that in the Indigenous interpretation of reciprocity, humans and animals are not separate, disconnected beings, but are believed to share 'spirit' and 'soul,' terms that signify an overlapping and enlargement of selves. As Indigenous scholar Doreen E. Martinez states, 'Your reciprocity must be collective rather than didactic (you and I), Christian (give and get), or capitalistic (receive and owe)' (Martinez 2021). It is also a challenge for students to grapple with the idea of land considered not as mere property to be owned and used as desired by humans, but as something that is sacred, alive, something to belong to. These ideas are, of course, impossible to truly understand without guided experience from an Indigenous adept, but elucidating them in the classroom does seem to make room at least for the *possibility* of an alternative conception of land and landscape. Reading further in Sakakibara's paper, we learn that in pre-colonial times, the Iñupiat traveled across the land on a seasonal basis, making homes in different locations, and learning intimately the ways of the land and its nonhuman inhabitants. The Iñupiaq creation story speaks of a whale that, in death, gave its body to form the land upon which the Iñupiat live, so the land itself is the gift of a whale, and therefore sacred. Since colonialism and modernity, and now, climate change, have accelerated the rate of change in the socio-cultural world of the Iñupiat, they are searching for and finding ways to adapt and still be Iñupiat.

An example that Sakakibara provides is the story of a village called Old Town, located in Point Hope, a promontory on the western coast of Arctic Alaska. The houses in Old Town were made of whalebones and sod – the multigenerational inhabitants were literally enclosed by the whale/land. When rising seas and coastal erosion caused much of Old Town to fall into the Arctic ocean, causing homes, ceremonial buildings, and meat cellars to disappear, the government resettled the people in New Town, a set of cookie-cutter North American single-family homes arranged on a grid. For the generation that experienced this change, it was a very painful adjustment. Now, according to Sakakibara, Elders in Point Hope tell stories about spirit beings that used to haunt the tundra but now live in the ruins of Old Town. These new stories, adaptations of traditional tales, help keep the memory of Old Town alive and inform young people who were born after the change. Here, too, we learn about storytelling as a cultural practice for adapting to change. The shared grief and solidarity across the generations help the community cope.

The whale hunt is a key part of the identity of the Iñupiat. The animal is thanked for the sacrifice of its life, and the meat is never sold; it can only be shared. The entire community takes part in the hunt, the dragging of the carcass over the ice, and the butchering and sharing of the meat. Stored in outdoor ice cellars, the meat can feed multiple families over the long, dark winter. Every part of the animal is used; to waste is to disrespect the sacrifice. The whale is hunted only because of need, not sport, and therefore, the community has traditionally imposed self-limitations on the whale harvest, taking only what need dictates. This notion of self-limitation is very common among Indigenous people the world over. Although environmental history tells us of inadvertent over-exploitation of natural resources by Indigenous people, including some animal extinctions, the track record of Indigenous people's successful co-management with their lands and resources over millennia stands in stark contrast to modern industrial civilization's rampage toward global catastrophe.

Now, climate change is causing bowhead whales to change their migration routes, and the melting sea ice makes it difficult to get to where the whales are.

Normally, the Iñupiaq whaling teams camp on the edge of the sea ice that may be two or three miles away from the shore. But with the ice melting and thinning, this has become a challenge, and often, the thin sea ice poses a danger. The rate of change of the environment is so fast that it is not always possible for traditional knowledge to keep up, as was the case in the story of The Scientist and the Elder.

The sole traditional musical instrument of the Iñupiat is the drum. Drums are made from whale parts, such as the stomach lining, and new ones need to be made every year. Drumming is an important part of the Iñupiaq tradition. Drums are played during celebrations of a successful whale harvest and during ceremonies throughout the year. With fewer whales, materials for making drums are harder to get, and there are fewer celebrations. The ceremonies that are held throughout the year are disrupted, and this causes a certain social disharmony and anxiety. Whereas whales used to provide material for the drums, now, drummers are inviting the whales to return to them. According to a quote in the paper, 'a whale is always looking for a village with good music.'

The intimacy of the Iñupiaq relationship with the whale depends on the viability of the sea ice. The sea ice has allowed the Iñupiat to live and thrive in one of the harshest environments on the planet. To help students understand this, we read, in separate small groups, extracts from a beautiful illustrated book called 'The Meaning of Ice,' co-authored by scientists, scholars, and Indigenous Elders (Shari Fox Gearheard et al. 2017). There are striking photographs of the terrain and the people and animals, and brief accounts by the people who live there. From this book, we learn that the Iñupiat think of the sea ice as 'home,' and while they don't live directly on the sea ice, it makes them feel homesick if they are away from it for too long. It allows for travel from village to village, and it leads them to where the Bowhead whale swims. Young people play sports on the ice. Elder Nancy Neakok Leavitt says that sea ice is crucial for the very existence of the Iñupiat, allowing a cashless, subsistence way of life, providing their bodies with the kind of food needed to survive in the cold. Other Elders describe the sea ice as being 'like a garden' (Shari Fox Gearheard et al. 2017). We learn that the Iñupiat can read the mood of the sea ice through its vast repertoire of sounds; indeed, the variety of sounds, shapes, and types is reflected in the complexity of the language, Iñupiaq.

Both these readings foreground the importance of stories in the context of Indigenous cultures: stories as ways to convey rich cultural and ecological information between generations, stories as teaching tools for ethical and responsible behavior, stories also as coping mechanisms for the catastrophic changes underway. As I've described in Chapter 5, stories also have an onto-epistemological role.

An additional short reading illuminates Indigenous value systems: an extract from Indigenous scholar and forest ecologist Robin Wall Kimmerer's celebrated book, *Braiding Sweetgrass: Indigenous Wisdom, Scientific Knowledge and the Teachings of Plants* (Kimmerer 2015). The extract describes a creature called the Windigo, part of the cultural traditions of many Native American tribes. The Windigo is a monster that is not born, but made, by the act of eating human flesh. Such a person transmutes into a creature ten feet tall, with limbs like tree trunks and carrion breath. Its tragedy is that the more it eats, the hungrier it gets; it is never satiated. Tormented by its inexhaustible hunger, it roams the winter landscape, seeking victims to devour. Originally a teaching tale warning of selfishness, greed, and cannibalism, the story of the Windigo has another, more modern interpretation. Students recognize immediately the destabilizing feedback loop implicit in a hunger that is *increased* by eating. It is not hard, therefore, to see the Windigo as a metaphor for Modern Industrial Civilization and neoliberal capitalism, with its endless hunger for resources, extraction, and growth. The discussion helps put in sharp relief the values of consumerist and individualist American mainstream culture.

Our study of Indigenous epistemologies has been made far more meaningful through guest lectures by Indigenous scholars and activists. For one class, Dr. Doreen E. Martinez, an Indigenous (Mescalero Apache) scholar and epistemologist at Colorado State University, gave a lecture on 'Racial Indigeneity and Climate Justice,' in which she spoke of Indigenous knowledge systems and the ethical insights they provide for engaging with the climate problem. She challenged us to not be passive recipients of the gift of Indigenous knowledge, but to take it forward into the world and into our lives. Present at the talk were representatives of my institution, including our Campus Sustainability coordinator, and indeed, the reverberations of that talk are still being felt. A year later, activist Kristen Wyman of the local Nipmuc tribe gave a talk about the attempts of the Nipmuc to reconnect with their traditions through food sovereignty, introducing to students the notion of 'rematriation' and the Land Back movement. Student responses to both these presentations were unambiguously positive, and for many students, revelatory. This made clear to me the difference between talking about Indigenous epistemologies and hearing from Indigenous peoples themselves about their experiences, ethical systems, and ways of knowing. There is no substitute for the latter experience.

If we look more closely at how the Iñupiat and other Indigenous peoples have related to the nonhuman species of their lands, we recognize in the concept of collaborative reciprocity the three meta-concepts of Balance/Imbalance, Critical Thresholds and Complex Interconnections. The idea that humans and nonhumans are on the same footing, and therefore equal, represents a kind of balance of power. The idea of reciprocity between equals also emphasizes a balance of shared responsibility and collaboration of life. Many Indigenous peoples refer to 'balance' as a central principle in their epistemologies, with individual balance extending seamlessly to a balanced relationship with the land and its nonhuman persons. For instance, the website First Nations Pedagogy Online (First Nations Pedagogy n.d.) states that 'individual balance is intricately bound to interconnection with one's immediate family, extended family, one's community and one's relationship with the world at large, both natural and spiritual.' In Gregory Cajete's book *Native Science: Natural Laws of Interdependence* (Cajete 2016), he speaks of the centrality of the concept of balance in Native American cultures, for example in Navajo cosmology. In classroom discussions, we explore how we might think about the term 'balance' in our lives, and how that can differ from the way the term is used by Indigenous people and also in the scientific context.

The meta-concept of Critical Thresholds is implicit in the way that Indigenous cultures self-impose limits on what they take from the land. For example, the Iñupiat never over-hunt the whale. The people of Parvati Devi's community in The Village Women and the Forest voluntarily impose restrictions on their use of forest resources so that the forest, including other species, can thrive. Collaborative reciprocity implies that to maintain a balance, one must restrain oneself from crossing certain thresholds, which are tantamount to breaking taboos. All of these ideas owe their existence to the recognition that all things are connected. This recognition of the inherent, a priori complex relationality of the world is perhaps the great common thread among Indigenous and forest-proximate cultures of the world. As Enrique Salmón states (Salmón 2000):

> Indigenous people view both themselves and nature as part of an extended ecological family that shares ancestry and origins. It is an awareness that life in any environment is viable only when humans view the life surrounding them as kin. The kin, or relatives, include all the natural elements of an ecosystem.

As I've briefly mentioned in Chapter 4, it is important to point out the difference between the way complex systems are viewed in science and scholarship, and the significance of complex relationality in Indigenous cultures. In the former, complex systems science emerges as a contrast to Newtonian mechanism, which assumes that the universe is machine-like, and complex relationality is regarded as a special case. Indigenous paradigms do not even need a term like 'complex system' because the universe is regarded as *a priori* complex, relational, and alive. In such a perspective, it is the simple systems that are the exception instead of the rule. Further, unlike the case of the average scientist of complex systems studying the field, where the subject-object separation is still dominant, in the Indigenous perspective, the observer *belongs* to the system that is being observed, participates in its doings, and has individual and collective responsibility toward it.

Thus, the three meta-concepts that we used to understand the basic science of climate change are already present, although with different specific meanings, in Indigenous cultures. Since they have multiple interpretations that become specific in particular contexts, these meta-concepts also qualify as boundary objects as described in Chapter 5. As they take on different meanings across disciplines, they also enable a diffractive reading across science and

Indigenous ways of knowing. Thus, students examining a diagram (Diagram 1) in a paper about Iñupiaq value systems compared with science (Barnhardt and Kawagley, n.d.) can see both differences and similarities and understand the importance of equitable collaborations between scientists and Native peoples in Alaska. I have not, as yet, attempted to introduce Barad's ideas or even the diffraction analogy to my first-year students explicitly; this is work in progress. However, by the middle of the semester, the discomfort that some freshmen initially describe with the transdisciplinary approach has dissipated, and they are able to use both stories and the meta-concepts to travel through disciplines and paradigms with varying but encouraging degrees of success.

As we approach the end of the course, we consider the relevance and importance of Indigenous peoples' role in climate mitigation. We read about the Swinomish people of the Northwestern United States, their climate action plan, and their broad and holistic approach to health. It becomes clear to the students that concepts like collaborative reciprocity, embedded in a deeply relational onto-epistemological paradigm, are eminently practical and realizable in the world, and contrast sharply with the mechanistic business-as-usual approach to sustainability and climate action. All this is key to our discussion of climate solutions, which I elaborate upon in Chapter 7.

Thus, our exploration of sea ice at the intersection of three inter-related systems takes us across disciplines, geographies, and scales. It becomes clearer to students that we live in worlds of interrelated human-biophysical systems. Our study of different kinds of causality and how small effects can ripple through the system to cause a large-scale shift helps students understand the nature of complex relationships. The three meta-concepts and the four dimensions of this pedagogical framework are made evident to students (the psychosocial action dimension is discussed in the next chapter).

Returning to the issue of injustice and power hierarchies, these are integrated throughout the course in the multiple ways that I have elaborated in Chapter 5. However, it is important to emphasize the foregrounding of the local in our discussion of justice and power. Much of the mainstream discourse on climate change is top-down, privileging the global scale and washing out the experiences, struggles, and predicaments of communities on the ground. Starting with the local context and then connecting to the global scale – a ground-up approach – goes some way toward redressing this imbalance. This is one motivation for the recent prominence given to the concept of 'clime' by environmental humanities scholars (I have contributed to a multidisciplinary discussion in this volume (Yü and Wouters 2023)).

6.6 Putting It All Together: Group Work and Projects

As detailed in Chapter 3, a classroom culture inspired by transformational learning, the growth mindset and a Natural Critical Learning environment are key elements of this pedagogy. These are especially important for a freshman seminar at a university like mine, where at least half the students are first-generation college goers and are increasingly diverse ethnically and racially. A key aspect of this classroom culture is small-group discussions and presentations. For example, I might start the class with a question and ask students already arranged in groups around large tables to have a preliminary discussion among themselves. We then engage with the question as a class. Similarly, I use in-class worksheets for students to practice material introduced in lecture. But the most successful use of group work is for students to learn in small groups and then present their findings to the class as a whole: that is, to practice being teachers. For instance, students might work in groups to present different aspects of basic climate science, or to brainstorm about an ethical question, or to use critical thinking about a proposed climate solution. The rubric for grading such group work (when it is graded) includes points for full participation by every member and assigned (rotating) roles for each person. A very successful recent small-group activity was an occasion for students to teach each other the three meta-concepts and the relevant science through the use of props. This particular class had what was perhaps an unusual proportion of students reluctant to enact physical processes, so after trying and failing to interest them in dramatizing the greenhouse effect, I adjusted my approach. I gave them props: a large plastic box containing small wooden blocks, meter sticks and a fulcrum, and a Jenga set. Assigned different topics, such as carbon cycle or sea ice mass balance/imbalance, or critical thresholds in ecosystems and the planet, or complex interconnections, students had to come up with creative ways to use the props to present and teach the concept to the rest of the class. Several students stated that the activity helped them understand the concept better and that they enjoyed teaching the rest of the class. On another occasion some years ago, we visited a major museum in a neighboring city, during which students observed and critiqued the museum's climate change exhibit. I summarized student comments and sent these to the museum, which responded with a letter appreciating the points the students had made and promising us that these would be considered for future exhibits. In a more recent class, students examined the U.S. National Climate Assessment draft for Alaska, which had been recently opened for public comments. Each group read different sections of the draft and then shared their comments with the class. I then forwarded the comments to the lead author of the Alaska chapter, Dr. Henry Huntington, who sent back a response thanking the students, appreciating specific points they had made, and promising to carry their ideas forward.

The last two descriptions of group work/projects above exemplify the pedagogical importance of taking the learning outside the classroom and making it matter in the world. Perhaps the most ambitious example of this was a pre-pandemic project making the connection between Arctic and global climate change and the changes happening in Massachusetts. Students read about climate projections for the state under various scenarios, and discovered that heat waves were a specific threat, with the very young and the elderly being especially vulnerable to heat illness. This is the case in part because Massachusetts has a generally cold climate, and most houses do not have (and have not needed, until recently) air-conditioning, and also because many elderly people live alone. Residents are unfamiliar with the symptoms of heat illness, which can progress very suddenly from mild discomfort to life-threatening heat stroke. We decided to create an infographic warning of the different stages of heat illness correlated with temperature and humidity for the elderly in our community. For this, students worked in small groups for about half the semester. First, they interviewed senior citizens who had been invited to an on-campus senior citizens' health fair by our nursing department. They asked them questions about the extent to which the elderly were aware of the impacts of climate change and the dangers of heat illness, and discovered that there was very little awareness of this potentially major problem. They asked them also what kind of information about this issue would be helpful to senior citizens. Having collected this information, we then had a guest lecture about heat illness by a professor of nursing. After further research, students created a number of infographic designs in small groups. We checked the information with two professors of anatomy and physiology. We then hired a student of graphic design to integrate the infographic prototypes into a final, professional design. Throughout this process, we were in conversation with our city's health department, which had not been fully aware of the risk from heat illness to the elderly population, and were very welcoming of the student efforts. The final design was then presented to the city health department.

Apart from its academic content, my freshman seminar on Arctic climate change is intended to foster college-level academic skills, familiarize students with various academic and non-academic support services on campus, help them feel a sense of belonging, and develop the kinds of meta-cognitive skills that are necessary for their holistic development. These are integrated into the academic content whenever possible. For example, students start to learn about systems by exploring the ways in which the university is a system, and understanding its physical infrastructure as well as its organizational arrangements. Another example has to do with learning good note-taking skills from lectures, articles, and videos, which are immediately applied to the material at hand. Short 'skills' assignments are interspersed with – and related to – the content. The book *Teach Students How to Learn: Strategies You Can Incorporate Into Any Course to Improve Student Metacognition, Study Skills, and Motivation* by Saundra Yauncy MacGuire (McGuire and Angelo 2015) has been invaluable for its many short exercises on research-backed learning strategies that help students develop the necessary metacognitive skills.

6.7 Challenges and Successes

Apart from inherent difficulties related to teaching climate change, there are specific challenges with teaching a course 'Climate Change and Social Justice in the Arctic and Beyond' to freshmen. One is the challenge of transdisciplinarity. Students are so used to the compartmentalization of education into watertight disciplinary boxes that they can be thrown into cognitive confusion in a course where scientific concepts and issues of justice and ethics might be discussed in the same session. The use of stories helps greatly in this regard. I have taught this course four times, and my earlier two attempts did not emphasize the use of stories as much as the last two. Stories make apparent the need for transdisciplinarity, and I emphasize again and again that in the real world, disciplines are always entangled.

However, the biggest challenge in this course has to do with classroom expectations in the transition from high school to college. This is in part because my standards are high. Incoming students are sometimes confounded by the greater degree of independent work and the higher intellectual expectations of college. Many of them are uncomfortable with failure, especially if they have sailed through high school experiences with an easy A. I have received the greatest pushback from students in all my courses, freshman or not, for imposing high standards. Of course, I also work extremely hard with students, often meeting with them one-on-one through much of the semester, to help them reach those standards. When they are able to take advantage of my help, the effect is transformative.

In all my classes, I emphasize the importance of high standards and express my faith in students that they can reach them with hard work, good study skills, and help. I demonstrate this faith by giving students multiple chances to succeed, without lowering standards, and also by spending as much time as we can both spare to help them get closer to realizing their potential. After the first exam, in which a failure rate of 50–60% is not uncommon, especially in my upper level physics classes, we discuss what both the students and I need to do better, and I show them a short video about the importance of the growth mindset. I emphasize the importance of failure as a teacher and give students examples of my own stumbles through my educational career. This is generally the point at which most students who are not doing well become open to trying harder. When they are able to be consistent with meetings and extra work, their turnaround is remarkable and inspiring. In one freshman seminar of 18 students, about 77% struggled academically in the first half of the semester, but 79% of these struggling students showed a significant turnaround by the end of the semester. Below are unedited extracts from several student reflections at the end of the semester, responding to a list of learning strategies we had worked on in class.

**Box 6.1   Freshman Seminar students reflect
on their learning at the end of the semester**

This class's learning from scholarly articles component was extremely helpful to me. This aided me in my other class, Composition II, because we were also learning how to obtain information for papers from scholarly articles. I was able to learn this more effectively by applying it in a subject other than composition II, which gave me stronger research skills for my essays/papers.

Prior to this class, I was unfamiliar with the term 'Indigenous.' I was aware of the concept but was unaware of this word. The lessons, articles, and videos we viewed in class, particularly the film 'Alaska Natives Fight for Their Land,' helped me grasp and comprehend the definition of the word as well as its significance. I also found the visit from someone representing Indigenous people and their challenges to be very informative and beneficial, as it helped me fully understand topics presented in class, such as the Iñupiat relationship with whales, and acquire new perspectives. However, I found the concept mapping idea to be challenging. I was able to understand the concept, but I feel it is something I will need to work on in order to improve.

This class has taught me many academic skills and academic knowledge I've never heard of. The skills I learned were thinking critically, ethically, holistically, concept-mapping, and Cornell method note-taking. In high school, I was able to do good without studying or taking 'good' notes.

The main things that I have adapted into my learning is accepting failure. During the semester, I had many hiccups along the way and found myself stressing over everything little thing as a result. Another thing that I learned is that I have learned how to be more of a quality writer over a quantity writer. I was able to develop skills that allowed me to focus on one specific thing and elaborate on that point. Having optional and make-up assignments also helped me get over these things, but they also gave me a way to relieve stress if I thought I did not do so well on a specific assignment.

I have definitely been able to learn better from scholarly papers in this class. From the different scientific papers we have learned from, I have learned to harness that knowledge and use it as my own. I believe I still need to work on learning from failure however, as I still feel very discouraged after seeing a bad grade.

I have learned a lot about open response questions in this class after not doing well in the beginning. Realizing that if we use Blooms taxonomy, then the answers I am giving will be so much more in depth and college-level responses. As well as learning about people working to stop planetary destruction like Henry Huntington and being able to affect the National Climate Assessment is very interesting and cool. This also helped me understand we live in a system and that everything is interconnected in different ways. Things we do can affect places or life in ways that we couldn't even imagine.

Concept mapping helped me a lot with learning the meta-concepts because you can see how all the ideas are connected and can correlate to each other which creates a huge web of information that makes it easier to understand as a student.

I have learned so much in this class! For example, my academic skills. You have taught me how to try my best at something and if I fail, to keep trying because no one gets things perfectly done the first time. I definitely need to work on not procrastinating or giving up if I fail at something.

So, I feel like this class more than any of my other classes really prepared me on how to be successful not only in the classroom but in the world. We went over so many important things that just really changed the way I looked at the world. I noticed in college it doesn't matter if you play a sport, you still need to do the work; professors won't treat you special because you play a sport. But going into the academic skills, I learn the main one was a note-taking method I tried the Cornel note-taking method, but it did not work, so I combined it with my high school note-taking method, and that work amazing. What I did was split the page and still did everything the Cornel method teaches us to do, but I also color code thing because this is how my brain works. But I feel what really change my learning is something you said to the class early in the semester, and that was reading over your notes right before class. This makes it so when you show up to class your brain is already working in the right direction and makes you prepared and ready for class. I also learned one thing that wasn't said in class or on paper but instead shown and I will be able to continuously use this throughout my learning journey, and this is showing your brain information in many ways. We did this in class by acting things out, reading notes, listening to lectures, and watching videos, and by doing this we give your brain a different point of view of this situation every time and by end your brain has a complete understanding of the topic.

Being in this class, I've learned that getting a good grade on assignments means that you will really have to try and show that you are learning the material in your answers. Two-sentence answers won't really cut it for this class or any class work, really. Dr. Vandanah has

> shown me that I need to connect the dots in my answers, and thoroughly explain my answers, and take my time with the assignments to get a good grade. I also learned that when walking into a class like this RAMS 101 class, you won't be able to just sit back and relax and watch your peers work. When in a class like Dr. Vandanah's, you need to be an active participant in your group and class, and I definitely learned that lesson well. A skill that I have learned about but am still working on is organizing my time correctly so I have enough time to work on my assignments and turn them in so they're not late. I still have trouble with this skill because being in college, there are not many people making sure you got all your assignments done except for you and potentially your friends so I'm really working on my accountability.

An additional challenge for a freshman class is that first year students are often distracted by the multiple learning curves of the college transition. The impact of the pandemic in recent times also seems to have – in my anecdotal experience – increased the severity and frequency of anxiety, family troubles, depression, and disengagement from learning. This has been very challenging for faculty, and burnout is a real danger. Therefore, I have had to negotiate a very difficult balance between maintaining high standards and giving students the help they need. However, in the latter half of the semester, the fruits of the hard work, individual help, and community building start to become apparent. The change in the freshman cohort is palpable. They are more serious, a little more confident, and able to engage with ideas that were initially difficult for them. The last time I taught this seminar, I had students write a final reflection paper. The metacognitive skills that they had developed by the end of the semester were apparent; some of them expressed amazement at the progress they had made.

There is another kind of challenge, which is onto-epistemological in nature, that arises when teaching climate change and social justice with reference to Indigenous experiences, predicaments, and ways of knowing in a context that is far removed from the latter. A typical Western classroom setting where the rest of Nature is literally walled off, where the emphasis is on the individual, where the learning is – despite my efforts to the contrary – mostly from texts rather than experiential – offers little space and much resistance to alternatives. The dominant paradigm is right here, in the classroom. I consider this problem in the final discussion in Chapter 10.

References

AJ+, dir. 2017. *This Is the Story of Alaska Natives' Fight for Their Land [Our Fight To Survive, Pt. 1]* | *AJ+*. https://www.youtube.com/watch?v=50_kse-Uh-g.

Ask a Biologist. 2014. "Frozen Life." Text. Ask a Biologist. July 9, 2014. https://aska-biologist.asu.edu/explore/frozen-life.

Barnhardt, Ray, and Angayuqaq Oscar Kawagley. n.d. "Indigenous Knowledge Systems/Alaska Native Ways of Knowing," 24.

Cajete, Gregory. 2016. *Native Science: Natural Laws of Interdependence.* 1st edition. Santa Fe, NM: Clear Light Publishers.

Chance, Norman. 2002. *The Iñupiat of Arctic Alaska: An Ethnography of Development.* Belmont, CA: Thomson Learning.

Clark, Douglas A., and D. Scott Slocombe. 2009. "Respect for Grizzly Bears: An Aboriginal Approach for Co-Existence and Resilience." *Ecology and Society* 14 (1). https://www.jstor.org/stable/26268040.

First Nations Pedagogy. n.d. "Holistic Balance Overview from the First Nations Pedagogy Online Project." Accessed April 16, 2023 https://firstnationspedagogy.ca/holistic.html.

Fisheries, NOAA. 2023. "Bowhead Whale | NOAA Fisheries." NOAA. Alaska. March 8, 2023. https://www.fisheries.noaa.gov/species/bowhead-whale.

Gearheard, Shari Fox, Lene Kielsen Holm, Henry Huntington, Joe Mello Leavitt, and Andrew R. Mahoney, eds. 2017. *The Meaning of Ice: People and Sea Ice in Three Arctic Communities.* Montreal: International Polar Institute.

Goosse, Hugues, Jennifer E. Kay, Kyle C. Armour, Alejandro Bodas-Salcedo, Helene Chepfer, David Docquier, and Alexandra Jonko, et al. 2018. "Quantifying Climate Feedbacks in Polar Regions." *Nature Communications* 9 (1): 1919. https://doi.org/10.1038/s41467-018-04173-0.

Grotzer, Tina A.. 2012. *Learning Causality in a Complex World: Understandings of Consequence.* Washington, DC: Rowman and Littlefield. https://rowman.com/ISBN/9781610488631/Learning-Causality-in-a-Complex-World-Understandings-of-Consequence.

Kimmerer, Robin Wall. 2015. *Braiding Sweetgrass: Indigenous Wisdom, Scientific Knowledge and the Teachings of Plants.* 1st Edition. Minneapolis, MN: Milkweed Editions.

Krupnik, Igor, Claudio Aporta, and Gita J. Laidler. 2010. "SIKU: International Polar Year Project #166 (An Overview)." In *SIKU: Knowing Our Ice: Documenting Inuit Sea Ice Knowledge and Use,* edited by Igor Krupnik, Claudio Aporta, Shari Gearheard, Gita J. Laidler, and Lene Kielsen Holm, 1–28. Dordrecht: Springer Netherlands. https://doi.org/10.1007/978-90-481-8587-0_1.

Martinez, Doreen E. 2021. "Storying Traditions, Lessons and Lives: Responsible and Grounded Indigenous Storying Ethics and Methods." *Genealogy* 5 (4): 84. https://doi.org/10.3390/genealogy5040084.

McGee, David. 2022. "Permafrost." MIT Climate Portal. August 2022 https://climate.mit.edu/explainers/permafrost.

McGuire, Saundra Yancy, and Thomas Angelo. 2015. *Teach Students How to Learn: Strategies You Can Incorporate Into Any Course to Improve Student Metacognition, Study Skills, and Motivation.* Sterling: Stylus Publishing.

NASA. 2019. "Positive Feedback – Arctic Albedo." Lesson Plans. My NASA Data. My NASA Data. September 23, 2019. https://mynasadata.larc.nasa.gov/lesson-plans/positive-feedback-arctic-albedo.

NASA Climate Change, dir. 2018. *Disappearing Arctic Sea Ice.* https://www.youtube.com/watch?v=hlVXOC6a3ME.

NSIDC. n.d. "Science of Sea Ice." National Snow and Ice Data Center. Accessed April 16, 2023. https://nsidc.org/learn/parts-cryosphere/sea-ice/science-sea-ice.

O'Connor, Joseph, and Ian McDermott. 1997. *The Art of Systems Thinking: Essential Skills for Creativity and Problem Solving*. London, UK: Thorsons.

PBS. n.d. "Iñupiaq Whale Hunt." PBS LearningMedia. Accessed April 16, 2023. https://www.pbslearningmedia.org/resource/echo07.sci.life.coast.eskimo/Iñupiaq-whale-hunt/.

Sakakibara, Chie. 2017. "People of the Whales: Climate Change and Cultural Resilience Among Iñupiat of Arctic Alaska." *Geographical Review* 107 (1): 159–84. https://doi.org/10.1111/j.1931-0846.2016.12219.x.

Salmón, Enrique. 2000. "Kincentric Ecology: Indigenous Perceptions of the Human-Nature Relationship." *Ecological Applications* 10 (5): 1327–32. https://doi.org/10.2307/2641288.

Whiteman, John P. 2018. "Out of Balance in the Arctic." *Science* 359 (6375): 514–15. https://doi.org/10.1126/science.aar6723.

Yü, Dan Smyer, and Jelle J. P. Wouters, eds. 2023. *Storying Multipolar Climes of the Himalaya, Andes and Arctic: Anthropocenic Climate and Shapeshifting Watery Lifeworlds*. 1st edition. Abingdon: Routledge.

7   Critical and Ethical Thinking about Climate Solutions

7.1   The Psychosocial Action Dimension: Emotions in the Classroom

The ultimate aim of meaningful climate education is to empower learners with knowledge, motivation, and inspiration to act effectively and ethically in the world so that we can avoid the worst impacts of catastrophic climate change and build diverse, just, thriving, ecologically harmonious societies within planetary and local thresholds. It is much too late to avoid climate change, which is already upon us and worsening, but, at this point in 2023, perhaps *catastrophic* climate change is still avoidable. (And even if this is not so, what choice do we have but to keep persisting?) As is the case for anyone who has some knowledge of the seriousness of the climate crisis, 'climate doomism,' a term popularized by eminent climate scientist Michael Mann in his book 'The New Climate War,' (Michael Mann 2020) is a very real trap, as I discovered when I first started integrating climate change into my physics classes more than a decade ago. Descending into hopelessness, surrendering agency to inevitable, looming horror, distracting oneself with other things – these are, perhaps, natural and logical reactions to learning about the crisis. Climate anxiety is a real issue, afflicting young children as well as college students and adults (climate scientists are especially vulnerable) ('This Is How Scientists Feel – Is This How You Feel?' n.d.; Tosin Thompson 2021; Verlie 2021). When epidemics of social isolation and loneliness already make life difficult for young people (Lipson et al. 2022), the additional emotional burden of climate hopelessness seems to be too much to bear. This is especially true of students at a small public university like mine, many of whom are first generation, struggling to balance long work hours with college. Several are immigrants from Latin American and African countries, and many have responsibilities to their siblings and other family members. To add climate despair to what these young people are already dealing with seems heartless at the very least. And yet, they need to know the truth.

DOI: 10.4324/9781003294443-7

Of course, climate doomism only helps the powerful, who have a vested interest in preserving the system that maintains their hegemony. So, how are we to deal with it, in and out of the classroom? Certainly, we don't want to make false promises, or peddle polyanna-ish dreams of a golden future, where technology – or some other magical entity – will solve the problem. We need to tell students the truth in all its complexity, but we owe it to them not to leave them in despair. In addition, we want learners of any age to feel informed and empowered enough to take wise and meaningful action on the polycrisis – but emotions such as hopelessness, despair and apathy, or anger, which is not channeled effectively, can block both intellectual and political engagement. How to negotiate the narrow path between telling the truth and falling into the trap of climate doomism?

There are four key approaches I have found useful in dealing with difficult emotions around climate change (see also Box 7.1).

1   Building community – engendering trust between teacher and students, and among students, is necessary for engaging with difficult topics. This requires a different classroom dynamic than the 'sage-on-a-stage' traditional power hierarchy, as I've described in Chapter 2. Multiple inspirations from transformative education theory, the psychology of the growth mindset, social-emotional learning, and the Natural Critical Learning Environment can help create a trust-engendering classroom culture. A key component I have described earlier is the educator's commitment to the intellectual and affective well-being of *every* student, while, at the same time, having enough group work and 'teach each other' moments to foster student-student connections. Other trust-building approaches that are notable include time-banking as described by educators in India and the United States (Pallavi Varma Patil and Melanie Bush 2022).

2   Scheduling space for talking about difficult emotions, and using art/literature/writing to share them. Ecologist Steve Running first proposed in 2007 that learning the truth about climate change can result in emotional trauma that might align with the Kubler-Ross Grief and Loss scale (Elisabeth Kübler-Ross 2014). The Kubler-Ross scale indicates five stages of grieving a personal loss: denial, anger, bargaining, depression, and acceptance. Progressing through difficult emotions toward acceptance of the reality of the loss is an indication of healing. In the climate context, we can get stuck at one stage or another – denial or apathy, for example, or bargaining, by telling ourselves that perhaps things won't be so bad – anything to avoid feeling the pain. Anger can be channeled usefully, but unacknowledged or unfocused anger may be unhelpful, even harmful. Climate policy expert Daphne Wysham, elaborating on Running's notion of climate grief (Wysham, Daphne 2012), speaks of a sixth stage: Doing the Work. 'This means taking courage from each other as we look this monster in the eye and fight side-by-side in the battle of a lifetime.'

Box 7.1   Freshman Seminar Students reflect on what gives them hope and a sense of empowerment

I found hope and empowerment for myself, society, and the planet through group work activities done in this class. We were able to brainstorm new ideas and build on each other's views and ideas. We also aided and encouraged one another when someone was having difficulty understanding a concept or coming up with an idea for one of our activities by providing simple explanations. For example, in our last group activity, we all brainstormed a variety of ideas and built on each other's thoughts to come up with a decent answer that included everyone's participation and work. I also felt empowered after learning that the suggestions we presented to Dr. Huntington as a class were beneficial and that he will utilize them to enhance ways of expressing national climate to the public. This helped me believe that even the smallest actions I take may make a difference in the world and assist others.

Throughout the course, I found hope when we went to visit the planetarium and learned how everything was connected. This really stood out to me because it was cool seeing how everything worked together to survive and that there's a whole different meaning behind life. Another thing I found hope in was learning about the Indigenous people and their stories. They faced lots of adversity but kept pushing through and were very thankful even though they didn't really have much. Another thing that gave me hope was Kristen Wyman's presentation, she talked about all the struggles she has to face in her everyday life being a Nipmuc Indian and even though she has all these struggles she still gets up and fights every day.

I think I have hope for myself to succeed like I hope for the rest of the world. I want myself to become better, like how I have throughout this class, working with classmates, and participating. I want the world to have that too. I believe that it would beneficial for everyone to have that type of feeling.

I find hope in people taking action to help the earth climate whether that be locally or on a bigger stage. The one video of young women exposing the big brand companies CEO was empowering. To see people brave enough to not only attack this big company but to also do it in front of many people is proof that we can fight these power companies. Then there were also people helping their local ecosystem. We watched a short video of people restoring their coastline ecosystem by restoring the population of clams and planting trees. To revive the ecosystem, they helped restore balance in the food web and then they also helped to reduce the amount of CO_2 emission by planting trees.

I hope to find empowerment from more classes and classrooms like this, filled with students with the similar drive to learn about climate change and Indigenous people, knowledge, and cultures. Being in a room with other like-minded students would give me comfort and feel empowered because it shows that we as the future might have a chance in fixing things. It is also up to younger people to help restore the earth's imbalance, so the more youth that try to learn about it, the better. The more young people who learn about climate change and fill in the jobs that will give them a space to make change is very important to me. I hope to be a politician one day, if possible, I will definitely be making policies to help our planet as much as we can.

One moment I had hope for myself, society, and the planet is when we were all in groups and we had demonstrated a specific action using either Jenga, a pendulum, or styrofoam balls. I chose this moment because in my eyes I was able to see that the whole class really took something from the class and was able to show that to the rest of the class on their own with little help. Another instance where some hope grew in me is when we reviewed Dr. Huntington's paper and sent some suggestions and how he actually listened to our feedback. I say this because for someone who's as established as Dr. Huntington to accept feedback from students like [us] who has just scratched the surface of social justice and climate change means a lot and makes us feel heard.

Personally, I found a lot of hope in the increased consideration of Indigenous knowledge in coming up with climate change solutions. As I learned in this course, Indigenous people often have a lot of knowledge on how things are connected and can come up with really effective solutions for combating the effects of climate change. They also have a greater respect for nature, seeing animals not as just resources but as relatives that are giving themselves to help them survive. These things make including Indigenous people in coming up with solutions for climate really important, so seeing that they are being included in these discussions make me hopeful for the future.

When we discuss the climate crisis, I inform students that difficult emotions may arise as we study it, and these are normal. I explain that acknowledging emotions is important for our mental health, and also because certain emotions can block us from meaningful action. I emphasize that the climate problem and the polycrisis are too overwhelming to face as lone individuals, which is why we must support each other as we study them. Introducing the six stages, I then make time to discuss how we feel according to these stages, and I share my own experience of painful emotions. I ask students to add other emotions that may not be on the list, and to include any complex emotions – such as feeling sadness and anger at the same time. Students have added feelings of fear, hopelessness, and confusion to the six stages. About four times in the semester, I use surveys – the six stages in six bubbles with space to add more, with a Likert scale for each stage – for students to examine the extent to which they feel different emotions at that point in their learning. This is a meta-cognitive activity that is helpful for students to increase their self-awareness. When we discuss these matters, students realize that they are not alone in feeling the way they do, and that we tend to cycle through different emotions and stages at different points. So, if we feel despair today, that doesn't mean that we have to stay stuck in that stage – we may feel ready to 'do the work' tomorrow.

I also share with students projects like Is This How You Feel, which records climate scientists' emotional reactions to their work. Probably no other profession is more intimately involved with recording and understanding the devastating changes underway on our planet. Decades of warning governments and the public have not led to meaningful action – it is no wonder that climate scientists are also vulnerable to despair and hopelessness ('This Is How Scientists Feel – Is This How You Feel?' n.d.). Sharing how scientists feel helps drive home the seriousness of the climate crisis, and also validates the students' emotional responses.

The humanities can provide much inspiration for negotiating grief. And, indeed, grieving is necessary – mourning the planet we will never have, grieving for the people who suffer and die as a result of climate extremes and social marginalization, and for the species threatened with extinction. But this can be overwhelming for the lone individual – such grief cannot be faced alone. When the classroom is a community, difficult emotions can be aired and shared without fear or embarrassment. One of the best ways I have found to 'walk through the valley of despair' and out to the other side is through poetry. On several occasions, I have invited a colleague, environmental poet Sam Witt, to facilitate a poetry workshop. Many students, especially in the sciences, are initially resistant – they don't think of themselves as potential poets and don't see the value of the exercise. But, despite one or two resistant students, the overwhelming majority of students change their minds after the exercise. When they read their poems out loud, their grief is held in community, validated, and acknowledged. I have personally felt the power of this exercise, after sharing my piece – an almost visceral feeling of being set free to think and act, at least for a while. When the stages of anger, despair, or depression re-appear, as inevitably they will, it helps to know that one need not be trapped in them, that there is room to walk through the darkness together

3 Studying the role of social movements, including youth and Indigenous movements. When we talk about climate solutions (as detailed later in this chapter) and the importance of moving beyond individual actions to collective action, it is helpful to look at the role of social movements in bringing about needed change. We know that social movements are key to applying pressure to governments and corporations – history abounds with examples from different contexts: the ozone problem to civil rights and anti-colonial struggles around the world. Therefore, I have found it useful to bring the role of social movements to student attention. Students work in small groups to research different kinds of climate-specific movements as well as broader movements of Indigenous resistance. Examples include 350.org, the Sunrise Movement, Extinction Rebellion, Fridays for Future, Mothers Out Front (local to Massachusetts), Indigenous Environmental Network, Idle No More, and Amazon Frontlines. Each group then presents to the class the key aims, scope, and accomplishments of these movements. I show students video clips of youth and Indigenous leaders as well as activist-scientists taking action and making speeches: Nemonte Nenquimo, Vanessa Nakate, Peter Kalmus, Greta Thunberg. In my freshman seminar on Arctic climate change, which allows me the full semester to teach about climate, we also look at the significance of Indigenous and other resistance through scholarly work. For example, reports from the Rights and Resources Initiative tell us that Indigenous people, through management of their lands over millennia, help sequester about 300,000 million tons of carbon (Rights and Resources Initiative 2018). Where Indigenous people have legal claim to their lands, the forest cover is thicker and the forests are healthier. Indigenous and other protests have stopped or stalled some 25% of fossil fuel and green energy projects (Temper et al. 2020). (The fact that Indigenous people must protest green energy projects as well as fossil fuel projects underlines how technological switching without system change can uphold the same colonialist, destructive mindset that brought us the climate crisis.) Another recent report from Oil Change International and the Indigenous Environmental Network (OCI Team 2021) estimates that Indigenous resistance the world over has kept about a quarter of US and Canada annual carbon emissions from the atmosphere

4 Building a sense of empowerment. Talking about resisting power structures feels abstract until students experience their own sense of empowerment. This is related to building community through regular group work where students help each other and teach the class. As mentioned in Chapter 2, the chance to offer their thoughts and concerns anonymously every week or two weeks, and to see that how they think and feel matters to the instructor, builds students' sense of self-worth and confidence. Multiple students have expressed to me, over the years, their appreciation for this practice. I also teach research-backed learning strategies, in class and in customized individual sessions. If students implement these, they can

make a dramatic difference in their performance. I repeatedly, and through every encounter, show students verbally and with microaffirmations that they matter, irrespective of how they are doing in the class. All of these enable students to feel valued and motivated.

Specific to climate change and the challenging emotions that often accompany it, my strategy has been to include a meaningful project for students to do in small groups. These can be as simple as, for instance, presenting short talks and slides on the basic science of climate change to an English lit. class studying climate change from a literary perspective, or more ambitious ones that I have described near the end of Chapter 6. During one semester, a calculus-based physics class explored the State Climate Adaptation plan, relating it to their future careers in engineering and medicine, and created displays for an annual Green Festival for which the state official who had led the climate adaptation planning team was the chief guest. Receiving his approbation and appreciation meant a lot to the students

Often, people ask me about hope. Surely, a good pedagogy must instill hope in the learner? This is not easy to answer because the polycrisis is so serious and there has been such little progress that hope can be conflated with an unrealistic, even delusional optimism. I have written elsewhere (Singh, Vandana 2021) about the problematic politics of positivity. But there is a kind of hope that arises from defiance, determination, love, commitment, and channeled rage that does not deny the hell that is coming or the hell that so many communities and species already face. It's the kind of hope that emerges despite – or because of   walking through the valley of grief and despair. When we walk with – to the extent possible – marginalized communities in their struggles for survival, dignity, and sovereignty, we don't just see suffering. We see creativity, courage, and other value systems and ways of knowing. These can be inspiring. Perhaps 'hope' is not the right word. Perhaps it is an 'optimism of the will,' as per Antonio Gramsci, or the discovery of meaning in the midst of struggle. Sometimes, it seems to me that to talk of hope is to reinstate despair. Because hope and despair can both be traps – if we depend on feeling hopeful in order to act, does that mean that when we lose a battle, say, to protect a rainforest, when we must give up hope of success, we will cease to act? Surely not!

The student voices in Box 7.1 indicate that hope arises in community, while doing activities to help each other learn; hope arises when contributing to something meaningful in the real world; hope arises when learning about those who stand up to power and learning about Indigenous ways of knowing. All of these are about connection and community. Many of these reflections indicate a kind of epistemic shift – hope arising when all hope seems to be lost, because of something you did together or learned together that shook your assumptions to their foundations. It is my sense that it is a mistake to focus exclusively on the individual experience of an epistemic shift, as seems to be the case in transformative education theory. My suspicion is that epistemic shifts rarely occur in isolation, but in the highly individualistic Western epistemologies, we tend to ignore the role of others. There is such a thing as collective intelligence that arises when there is community, trust, caring, and a nonlinear coalition of minds.

If hope – or 'optimism of the will' – is necessary, then we must believe that better futures are possible. Before we recognize the hold that the dominant paradigm has on our daily lives, our survival, and our imaginations, it can be very difficult to conceptualize and work toward futures where we are no longer under the sway of corporate colonialism, where ecocide and species extinction have ceased, where social hierarchies are being flattened, and where we have figured out how to make and live in diverse social-ecological cultures. But this re-imagining is necessary. This future sense that develops through the imagining of alternatives can allow us to inhabit, however briefly, a different world, a better world.

In order to hone this futures skill, however, we must take a critical look at current-day purported climate solutions. We live in an age of massive greenwashing by the powerful (Carrington 2021). It is a mistake to forget, for even a moment, the constant presence of power when we think about the polycrisis. Power hierarchies drive the polycrisis, of course, but the powerful also *use* disasters, confusion, and chaos to consolidate and expand their power and control. Not only does power corrupt, it also co-opts. Carbon Market Watch and New Climate Institute have produced a recent report on corporate climate action claims, about which, they state (Carbon Market Watch 2022): 'Despite claiming to be champions of climate action, two dozen of the world's largest and richest corporations are hiding their climate inaction behind the fig leaf of green-sounding "net zero" plans.' An online project called Hoodwinked in the Hothouse: Resisting False Solutions to Climate Change (*Hoodwinked in the Hothouse* 2021) authored by a grassroots coalition of Indigenous and environmental justice groups analyzes a number of purported mainstream solutions and declares in its 2021 zine that 'As you turn these pages, you will enter a Pandora's box of climate false solutions, primarily designed to profiteer from the global ecological crisis. Most of these can be characterized as unproven techno-fixes, negative emissions technologies, carbon pricing mechanisms, corporate snake-oil products or extreme energy projects. All claim to address the climate crisis while avoiding the very underlying drivers that got us into this mess in the first place: economies of greed and hoarding; endless growth; corporate enclosure of land; erosion of biodiversity and the exploitation of life.'

7.2   Critical and Ethical Thinking about Climate Solutions

Given such governmental and corporate sleights of hand, how are we to distinguish real solutions from false solutions? How can we help students think critically and ethically about climate solutions? Through my years of teaching climate change and from reading of the experiences of others, I have come across a number of traps that prevent us from conceiving of real solutions. A partial lexicon is below:

1 Climate doomism – such paralyzing, overwhelming despair about the climate problem that we relinquish hope, agency, and action and turn away from the problem or decide to party until the bitter end. As I mentioned above, this only serves to leave an open field for business-as-usual

2 The Technofix trap – assuming that technology, including future tech, will solve everything. This comes from a perspective that sees climate change as a solely technological issue, practices carbon reductionism, ignores the polycrisis, and typically also ignores issues of justice and power (see climate scientist Michael Mann's interview here for his critique of the technocratic approach that ignores the politics of climate change; Watts 2021)

3 The Marketfix trap – the assumption that market mechanisms will fix everything. While it may be the case that market mechanisms, properly regulated and enforced, might, in the short term, help us change direction, it is unlikely that the same framework that caused the problem will be helpful in the long run. Often, people who live in this space tunnel between it and the technofix trap

4 The Individual Actions trap – assuming that our responsibilities end once we've got solar panels on the roof and an electric car in the driveway. Individual actions are valuable in their own right for other reasons, but unless they are connected to and influence collective networks, they are not going to help us with the climate problem, much less the polycrisis

5 The epistemological trap – this is what I've called paradigm blindness in Chapter 4. It assumes that the same way of thinking, acting, and living in the world that we are used to (the Dominant Social Paradigm of neoliberal societies and the underlying Newtonian mechanistic way of conceptualizing the world) is the default, the best way and the only way, and that a sustainable future will simply be a nicer, kinder extension of today. This is, of course, related to the Marketfix and Technofix traps

6 The privilege trap – assuming that your unacknowledged societal privilege whether of gender, race, national origin, or other – entitles you to be the one to solve the problem, including dictating to others how they should live. Because the privileged benefit from the current socioeconomic system, they are likely to cohabit the epistemological trap. An extreme form of this is what I call the megalomaniac trap; those in power are especially susceptible. This is the assumption that if you have enough money and power, you are especially qualified to solve the world's problems even if you or your company has contributed significantly to creating them

The pedagogical framework I have described in this book suggests a pathway for thinking about climate solutions. First, the science clearly implies the following:

- We need to cut carbon emissions so as to restore balance in the carbon cycle
- We need to restore the other violated planetary boundaries since all these thresholds are intimately connected and transgressing even one can imperil human well-being and that if the biosphere
- We need to do all of these things while promoting social equity and justice. This is important not just for ethical reasons but also for pragmatic ones. If large areas in the Global South become uninhabitable due to climate change, this will set off mass migrations northwards of millions of people, the like of which we have never seen. They will overwhelm infrastructure and capabilities of even the most well-off countries, assuming they are received without violence

Therefore, *a proposed climate solution must first be subject to the following criteria* before it can be deserving of further consideration.

1 Does it move toward restoring the carbon cycle by lowering carbon emissions?
2 Does it move toward restoration of violated planetary boundaries and respect all these thresholds?
3 Does it foreground and promote equity and justice, including the acknowledgment of justice as an intersectional issue, the existence of power hierarchies, the enduring legacies of colonialism? Does it, therefore, challenge power? Does it consider multispecies justice? (see for instance Menton et al., 2020)
4 Does it have non-linear efficacy? Does it scale out? Can it be integrated with other solutions? (I use the term 'scale out' rather than 'scale up,' as the latter implies a one-size-fits-all homogeneity. 'Scale out' as I've heard it used by activists such as Ashish Kothari implies adaptability to local conditions as well as local control.)
5 What's the paradigm? That is, what worldview, framework, and value system is the solution emerging from? Is it the Newtonian/Dominant Social Paradigm of Modern Industrial Civilization, which has resulted in the polycrisis we face today? If so, is it likely to work in the long term?

This framework helps us think critically about a proposed climate solution. If the solution does not pass the test with regard to these five questions – or if it generates more questions for further investigation – we ask ourselves under what conditions such a solution *might* become a real solution. I will illustrate by means of three examples.

Our first example is the Electric Vehicle (EV). Electric vehicles are touted as climate solutions because they use batteries that generate electricity to run the car instead of using gasoline. It is expected, according to some estimates, that there will be over 300 million EVs on the roads by 2030 (International Energy Agency 2022). Let us see if replacing gasoline-powered vehicles with EVs passes the test of the five questions above.

Will EVs lower carbon emissions? Perhaps not during manufacture, if the manufacturing plant runs on fossil fuels. It also depends on the source of electricity for charging the batteries. If electricity for both manufacturing and charging comes from wind or solar, then we can say that through manufacturing and use, EVs ought to lower the carbon footprint. Therefore, under these conditions, EVs pass the test of the first question.

What about other planetary boundaries? At this point, I introduce two aspects of the EV solution from a paper I co-authored with climate physicist Chirag Dhara (Dhara and Singh 2021). One, EV batteries require certain minerals, such as lithium ore. Mining for these minerals is expected to boom, to the extent that regulations are being put in place to mine the deep ocean, among the few ecosystems on Earth still free of direct human exploration. In addition, EVs need other materials like metals and plastic. A useful measure in addition to the carbon footprint is the *material footprint* of a product: the sum total of all the mass of materials needed for that product to exist, from extraction to manufacture, use, and disposal. Thus, in addition to mining for minerals for batteries, mining for metals would also increase. The damage to ocean and terrestrial ecosystems would be tremendous. Extractive activities are also sources of greenhouse gases, so they may potentially increase the carbon footprint of these vehicles. We are in the sixth mass extinction on our planet, this one induced by Modern Industrial Civilization, and it is not hard to understand that large-scale mining would worsen species extinction, especially in our already suffering oceans.

One other aspect needs to be considered: the mainstream economics notion of endless economic growth. If demand rises exponentially, it doesn't matter how carbon-neutral the vehicle is: its material footprint will also grow exponentially. In fact, the global material footprint is increasing at an alarming rate (United Nations n.d.). Developed countries that claim that they have decoupled their Gross Domestic Product from carbon emissions are generally ignoring their rising material footprints, and the offshoring of much of extraction and manufacture to the Global South (Wiedmann et al. 2015; Hickel et al. 2022).

Therefore, the EV solution is likely to violate other planetary boundaries. It fails Question 2.

What about question 3? Students point out problems with the affordability of electric vehicles. If only the rich can afford them, how is this solution equitable? There is also the fact that extraction and mining often take place in Indigenous lands in the Global South, from which these peoples are often violently displaced. Clearly, under these circumstances, justice and equity are not served. Nor are power hierarchies confronted.

Question 4 needs a little elucidation. Can EVs have a wide ripple effect? Pushed by policy changes and consumer awareness about climate change, certainly, EVs are poised to take over the transport sector. In the future, gas stations will give way to charging stations. Is it possible for EVs to hook up with other solutions for nonlinear change? The question is open. We leave that possibility for later and turn to Question 5.

Question 5 asks us to think about paradigms and systems. And, in fact, it becomes clear that the solution of simply replacing gasoline-powered vehicles with EVs does not change the status quo. The power hierarchy doesn't change. The ways of thinking, living, and being can stay the same. While former fossil fuel industry employees will need retraining for green energy jobs and fossil fuel magnates might suffer temporarily, the great engine of endless growth, the alienation with the rest of nature, the pyramid of inequality will be undisturbed, and may even become even more powerful.

I emphasize to students that we do need to switch completely to electric vehicles. There is no doubt about that. But by now it is clear to them that a mere *replacement* of one technology with another will not work. If the 'solution' violates other planetary boundaries, worsens inequalities and power hierarchies, and continues to perpetrate injustice and oppression, it is a false solution. So, under what circumstances might EVs become a real solution?

Students now understand that it is not enough to switch technologies; the *system* has to change as well. Therefore, what do we need to do in order that we use fewer cars so that we need not mine and manufacture as much? Good public transport immediately comes to mind. Imagine a public transport system that serves communities, is affordable, and entirely powered by electricity from renewable resources like solar and wind. Great, I tell my students. That's the low-hanging fruit. What else? Ideas start to come up, from car-pooling to city redesign. A discussion about alternative economics ensues. How do you ensure that everyone, including former fossil fuel industry workers, is taken care of? There are more questions than answers, but students are already widening their thought processes beyond EVs to see how they might connect up with other concerns, thereby engaging with Question 4. This rich discussion illustrates to students the crucial importance of looking at technologies in their wider socio-economic-political-ecological contexts rather than in isolation.

Another example of a proposed solution that has also given rise to rich discussions is the Half-Earth Project of the E.O. Wilson Biodiversity Foundation (E.O. Wilson Biodiversity Foundation n.d.), an ambitious plan to restore and protect Earth's biodiversity. The website states that if 'we protect half the global surface, the fraction of species protected will be 85%, or more. At one-half and above, life on Earth enters the safe zone.' Based on mathematical and ecological arguments, this sets a 'safe' threshold for biosphere integrity, one of the violated planetary boundaries.

An encouraging aspect of the Half-Earth idea is that it is a plan to restore and protect biodiversity – that is, natural habitats not limited to forests, including grasslands, ocean ecosystems such as mangrove and seagrass habitats. It is therefore far more sophisticated than the oft-repeated 'solution' of 'plant more trees.' Focusing on trees rather than forest ecosystems and more widely, other natural habitats, is dangerously simplistic because it can be easily co-opted by governments and corporates, allowing them to 'compensate' for destroying natural habitat by planting trees (which may be non-native) somewhere else. A plantation is not a forest, nor are ecosystems interchangeable. An extreme example of this is the Indian government's recent proposal for destroying a biosphere reserve on a tropical island for the purpose of building a transnational port – the proposal aims to compensate for the 'loss of a million trees' by planting 2000 trees in the relatively arid state of Haryana more than 2,500 km away (Pankaj Sekhsaria 2023).

Let us see whether the Half-Earth idea passes the 5-question test.

First, it is the case that increasing carbon sinks by restoring natural habitat is likely to pull some carbon dioxide out of the atmosphere, thereby lowering carbon emissions. With regard to the second question about other planetary boundaries, certainly this proposal seeks to restore biosphere integrity. It is possible that this may aid other planetary boundaries as well, such as land system change, since restoring natural habitat will mean reversing in part the current trend of taking over natural habitat for 'human use.' A healthier ecosystem might help species resist the effects of pollutants such as plastics (the fifth broken planetary boundary) although it does not directly address this threshold. But it does seem that this proposal can get part of the way through Question 2.

What about equity and justice? This is where the proposal falls short. Students have asked: will people – local and Indigenous – be displaced from their lands for the purpose of protecting those regions? Indigenous protests against the UN REDD project, which seeks to pay countries (in the Global South in particular) to *not* cut down their forests, point out that such schemes can become excuses for land grab, threatening the rights of Indigenous peoples to manage and live off their lands. A glance at the Half Earth website indicates that the Council for the project consists primarily of scientists from the Western hemisphere, and there is only one representative of Indigenous peoples, the Potawatomi scientist and writer Robin Kimmerer. There is also no indication of the awareness of Indigenous concerns, or of the success of Indigenous peoples' management of carbon and biodiversity worldwide (Rights and Resources Initiative 2018). So, at least from the website as accessed in March, 2023, there does not seem to be overt acknowledgment that such a proposal might be unjust and inequitable.

With regard to Question 4, it seems likely that an ambitious attempt to restore global biodiversity can only happen if such a project hooks up with other simultaneous efforts, such as reducing plastic pollution, using agricultural lands in a more ecological manner than the extreme methods employed by Industrial agriculture (agroecology appears to be a promising approach), and changing fundamental concepts of city design and transportation (for instance, designing around migration routes and natural habitats instead of destroying them). Unfortunately, there is little acknowledgment of the possibility of such interconnections on the website. In fact, it seems to be a common feature of proposed solutions that they are proposed in isolation from other possible solutions.

Finally, let us consider Question 5. It is clear to students that the Half-Earth solution as stated appears to assume a Human-Nature separation. This stands in contrast to, for example, the idea of collaborative reciprocity that is fundamental to Indigenous cultures (see Chapter 6). Therefore, this solution ignores the epistemological dimension and suffers from paradigm blindness. A paradigm of Western notions of wildlife conservation is indeed that Nature and Humans are separate – for example, American National Parks were created on Indigenous land by dispossessing Indigenous peoples, despite their co-existence on these lands for millennia. It is not that we don't need some areas set aside primarily for nonhumans; in fact, many local and Indigenous peoples do practice self-limitation and demarcate regions to stay mostly untouched – for example, the tradition of sacred groves in India (Madhav Gadgil 2018). But Western-style elite conservation methods have often violently dispossessed the people living in those regions (Rangarajan and Shahabuddin 2006), and in some cases, this has led to the deterioration or denudation of those lands.

Students also ask, at this juncture in our discussion, what the 'human' half of Half-Earth might look like. Does it mean a tacit acceptance of the current status quo, with many cities suffering from pollution and requiring enormous amounts of water and energy? As some Indigenous commentators and others have pointed out, why Half-Earth? Why not Whole-Earth? Why not fundamentally rebuild our relationship with the rest of Nature by discarding the dominant paradigm of Modern Industrial Civilization? (Kothari 2021). Of course, what stands in the way is power. And, like many proposed solutions, this one also ignores the prevalence of power hierarchies as both preventers and co-opters of solutions, and preservers of the current paradigm. As it is currently presented on the website, the Half-Earth proposal fails Question 5.

Now we turn to the question: under what circumstances might Half-Earth be transformed into a real solution? Ideas flow through the classroom. Here are some of them.

• Start with a Council of Indigenous peoples that truly represents their concerns and predicaments. Form a collaborative, equitable coalition with scientists. Understand privilege and power differences and co-construct a paradigm that respects relationships among and between human societies and the nonhuman world. Then draw out a plan that respects Indigenous land rights (already proposed as a climate solution) as well as biodiversity integrity. Make this a Whole Earth plan that also looks at urban and peri-urban social-ecological systems

- Connect with other necessary efforts to restore other planetary boundaries, which also means acknowledging and confronting corporate and state power – including ways to reign in the plastics and agricultural industries

My final example of a proposed climate solution is taken from Project Drawdown at https://drawdown.org/, probably the largest repository of solutions on the web. One of the proposed solutions we have discussed under the sector heading of Health and Education: Family Planning and Education. The relevance of health, access to family planning, and universal 'quality education' to climate action is explained thus (sans links):

> When levels of education rise (in particular for girls and young women); knowledge of, access to, and use of contraception increase; and women's political, social, and economic empowerment expand, fertility typically falls. Across the world and over time, this impacts population at a global level. Education, particularly around science, technology, engineering, mathematics, climate solutions, and sustainable consumption can help to strengthen the capacity of communities to support the growth of jobs centered on low-carbon technology and mitigate future emissions.

With some fairly nuanced discussion about population and consumption, including an acknowledgment of the often racist and classist history of population policy, the website goes on to further specify that access to and support for voluntary family planning, along with education for women and girls, is necessary for gender equality and also contributes to climate adaptation and resilience. I tell my students that there is a body of research to support the idea that the education of women and girls can be transformative not only for themselves and their families but also for their communities. According to UNICEF ('Girls' Education | UNICEF' n.d.):

> Investing in girls' education transforms communities, countries and the entire world. Girls who receive an education are less likely to marry young and more likely to lead healthy, productive lives. They earn higher incomes, participate in the decisions that most affect them, and build better futures for themselves and their families. Girls' education strengthens economies and reduces inequality. It contributes to more stable, resilient societies that give all individuals – including boys and men – the opportunity to fulfil their potential.

Let us subject this solution to the 5-Question test.

First, does this solution lower the carbon footprint? In terms of slowing population growth, possibly. But we cannot look at population in isolation from consumption, as the Drawdown website acknowledges. The vast bulk of carbon emissions comes from low-population Western societies, including some with very high educational levels for both women and men. It is true that we cannot have a runaway human population on a finite planet, but currently the global material footprint is rising at a higher rate than population growth (United Nations n.d.). Students also point out the troubling lack of specificity in the term 'quality education.' Yes, we need education to transition to a renewable energy economy, but, as per the thesis of this book, the *kind* of education is absolutely crucial. Mainstream education tends to maintain the status quo rather than to challenge it. This lack of specificity is also prevalent in descriptions of other purported solutions, as mentioned above. And yet, the devil is in the details.

So, if this solution doesn't entirely pass the first test, let's go on to the next one. Does it move us toward restoring other planetary boundaries? If it succeeds in lowering population growth, that is possible, but not likely if the rise in global material footprint continues to outpace the rate of population growth. Fewer people may not necessarily imply a lighter ecological footprint, considering the fact that the United States, at 5% of the world's population, consumes more than 25% of its resources.

With regard to equity and justice, this solution does much better. There is no doubt that the empowerment of women and girls is necessary for social justice and human rights. We live in patriarchal societies that daily inflict violence on women and girls. The education of women and girls has tremendous ripple effects, as the above-mentioned UNICEF site maintains. But questions remain as to whether rising income – which is a good and desirable result – may lead to a high-consumption lifestyle that can contribute to the violation of multiple planetary boundaries. As discussed earlier, this violation can result in species loss and the displacement of Indigenous peoples, perpetrating other forms of injustice. Again, the necessity of acknowledging the failure of mainstream education and the consequent need for radical transformation in education are conspicuous by their absence.

Family Planning and Education of women and girls are already acknowledged as having multi-directional ripple effects, so this solution doesn't fail Question 4, except in the ways discussed above. It is when we get to Question 5 that the issues become clear. It is not simply that women should be 'allowed' the freedoms and access to jobs, etc. to stand shoulder to shoulder with men. It is fundamentally the heart of the issue that patriarchy, which includes a violent assault on the rest of Nature as well as women, needs to be dismantled. Often, racial and gender justice are cast as merely problems of access and representation. These aspects are true, but only the tip of the iceberg. Letting women into positions of power in a system that is destructive to planet and people is not true gender equality. Therefore, a critique of the paradigm is necessary, both as it refers to gender and as it refers to education and the environment.

In the form described on the website, this solution does not pass the test of Question 5. But there are so many good things about this solution that it deserves re-consideration. I ask the students: how might we re-imagine this solution so that it can be a real solution?

Central to the ideas that emerge from the discussion is the need for transformative education – one that is based on a paradigm of complex relationality, that teaches life skills and working together as much as formal knowledge, one that is transdisciplinary, oriented toward working with communities and learning from them, and one where the study of gender and sexuality honors the rights of women, gender-fluid persons and people of all sexual orientations. A key aspect of this education must be ecological literacy and practical skills in caring for and restoring damaged ecosystems, including skills that encourage communities' food sovereignty through, for example, community gardens. Integrating Indigenous knowledge systems into education by building egalitarian and respectful relationships with local Indigenous groups can be mutually beneficial. Climate change education in particular should go beyond science and technology to concerns of justice and power, and include ways to make meaningful change from beyond the four walls of the classroom.

In my experience, the use of the 5-Question test for proposed climate solutions has been immensely useful in enabling students to think critically, ethically, and holistically about climate action, as the above three examples illustrate. These exercises help them develop a healthy skepticism about, for example, the corporate scramble for 'net zero'; some of the critiques from students of carbon offset schemes, for instance, have been scathing. But the 5-Question framework doesn't just help with critiquing solutions, it provides a methodology for understanding how some solutions that fail the test might be salvageable and useful when emplaced and modified via a different socio-economic-ecological context.

The holistic or complex thinking skills of students can be initiated by these exercises because they can see that solutions must connect with each other to work, and also because it helps them think beyond narrow categories. It becomes evident that we need, simultaneously, multiple kinds of change to engage usefully with the polycrisis. Starting from the need for a paradigm shift, the emergence of a new paradigm implies the following kinds of simultaneous and interconnected changes:

- Technological transformation – green energy and energy savings for all
- Sociological change – shift from individualistic and isolated ways of thinking and living to a different set of values that allow for participatory democracy, power to communities and renewed relationships with the rest of Nature
- New legal and governance frameworks – to 'make polluters pay' and shift funds to the green energy economy; to make ecocide a crime; to allow

rights to indigenous people over traditional lands; to protect those most vulnerable; to shift from colonialist top-down governance to make space for participatory democracy

- Changes in human living arrangements and agriculture – redesigning cities via ecological and just design principles; agroecology; farm towers for feeding large populations healthy plant-based diets
- Alternative economics – economic systems that recognize and respect natural critical thresholds such as planetary boundaries and meet the needs of all, that is, regenerative and distributive economics; sustainable energy usage
- Ecological transformations – rewilding and restoring, with the leadership of and collaboration with Indigenous people and others – healing ecosystems on land and in the ocean so other species can survive and help restore the web of life

The Swinomish Indian Tribal Community is an Indigenous group in Washington State in the Northwest United States that has come up with a much-praised Climate Action Plan. A Washington Post article provides details (Morrison, Jim 2020); we complement this reading with a video about the restoration of traditional clam gardens (Swinomish Indian Tribal Community 2021) and a very brief look at the tribe's own document (Swinomish Indian Tribal Community n.d.) on their Climate Initiative.

What the whale is to the Iñupiat, the salmon is to the Swinomish. But warming waters due to climate change are affecting this temperature-sensitive species, as is habitat destruction through the draining of tidelands and building of sea walls. The Swinomish are working on restoration of the tidelands and are planting trees along streams to cool the waters.

Jamie Donatotu, tribal environmental health officer, is quoted in the Washington Post article. 'The salmon and the crabs and the clams are relatives. They're living relatives. They're not just resources. And so you treat them with a symbiotic respect. They feed you because you take care of them. It's a very different way of thinking about why these areas are important.'

The Swinomish are also bringing back traditional practices such as clam gardens and oyster beds, which revitalize local ecosystems, provide some protection from sea level rise, and bring back food security to the tribe, while also connecting them with their traditions. 'Once in place, the garden will create a spot for elders to share stories, passing on tribal knowledge. It will supply a first food while serving as an example of the tribe's resilience, self-determination and cultural stewardship, all health indicators.'

In fact, the Swinomish have a very broad definition of health, not limited to physical health of a single individual, but extending holistically to the well-being of the community, including nonhuman relatives, and all of the environment. In the assignment that follows these readings, students are asked to describe how previously encountered concepts like collaborative reciprocity and cultural resilience are demonstrated in the work of the Swinomish – and, indeed, the action plan makes lucid these concepts. Further, I ask students to reflect on how the three meta-concepts manifest at the local scale in this context. Finally, we examine the Swinomish climate action plan through the 5-question framework. Assuming it is carried out as indicated, is it a real solution or a false solution?

The restoration of ecosystems is recognized widely as a climate mitigation tool (UNEP/FAO 2021) that lowers carbon emissions – therefore, this solution passes the first test, as it is oriented toward restoring the carbon cycle. With regard to question 2 about other planetary boundaries, biosphere integrity (at the local scale) is at least one other boundary that is addressed here. Because chemical fertilizers are key to conventional agriculture, if local, traditional foods displace conventional foods, this would not only reduce carbon emissions but also address locally the nitrogen cycle imbalance. The Indigenous practice of mixing old, broken clam shells into the beach may address ocean acidification under certain conditions at the local scale (Jen Schmidt 2022). Turning to question 3 on equity and justice: a solution that builds back the dignity and sovereignty of an Indigenous tribe against centuries of settler colonialism, and which also acknowledges the right of other species to exist and thrive, passes this test. Question 4 on how this solution might scale out and connect with other solutions allows us to muse on the possibility of restoring ecosystems and local foods in other spaces too. We ponder whether the adoption of renewable energy infrastructure as well as energy conservation measures might enable the tribe to go even further in their climate action plans. The fact that this solution is not just about climate change but also acknowledges the connection to health and its multiple dimensions indicates that it may very well be a *nexus solution*, addressing multiple issues at once. Finally, it is clear that this solution emerges from a paradigm that is founded on connection and relationality, countering the hegemony of the Newtonian mechanistic worldview.

So far, I have talked about climate solutions mostly as mitigation strategies. Climate adaptation (adapting to changes underway or inevitable) and mitigation (making the climate problem better, or at least preventing it from getting worse by controlling emissions) are often taken to be two entirely different and independent things. However, the Swinomish Climate Adaptation plan has aspects of mitigation as well – restoring biodiversity and moving us toward carbon cycle balance. Parvati Devi and her fellow village women are simultaneously adapting to climate change and contributing to its mitigation by restoring their forests. Perhaps it is time to consider adaptation and mitigation more holistically, and here again the 5-question framework and variations thereof might be useful.

Relevant to this discussion is the important concept of maladaptation: 'Broadly defined, maladaptation is when climate change adaptation actions backfire and have the opposite of the intended effect – increasing vulnerability rather than decreasing it,' say researchers in a post discussing their work

(Lisa Schipper et al. 2021). External adaptation aid can further entrench old inequalities and power relations, and can create new vulnerabilities for communities. Thus, for multiple reasons, it is wise to subject both climate mitigation and adaptation strategies to critical and ethical scrutiny. Adapting the 5-question framework to address both mitigation and adaptation is a work in progress.

7.3   Speculating the Future

One aspect of the Newtonian paradigm as a socio-scientific paradigm beyond physics is its fragmentation of space, time, and relationships, as suggested by Table 2.1 in Chapter 2. Short-termism is a specific kind of temporal fragmentation, where we are simultaneously disconnected from the past and unable to think about the future in the long term. We can see this in such short-term phenomena as election cycles and quarterly and annual reports. At the same time, many of us cannot fully inhabit the present, because we are constantly anxious about the near future. This is especially true of those who are in the bottom third or so of the pyramid of power in Figure 5.3. The removal of the social welfare safety net has brought precarity into the lives of working people. I can see this phenomenon in my students, many of whom are troubled by combinations of student debt and having to work long hours at tedious jobs to pay expenses, while often being responsible for siblings or older family members. It is very difficult to think about the medium-term or long-term future when you are worried about the next paycheck. These pressures on the young, along with the fragmentation of familial and community networks are likely responsible for, or at least exacerbate, the depression and anxiety crisis among the young in the United States. By contrast, the wealthy have reserves to draw upon in 'difficult' times. It is perhaps not a surprise that the area of futures studies or Foresight studies has been dominated by the rich and powerful, in the shape of corporations and governments of developed nations, with academia arriving somewhat late to the discussion.

Philosophers and observers have commented on the colonization of the future (Krznaric 2019): just as early European colonizers conceived of new (to them) lands as *terra nullius* or *empty land* erasing the long occupancy, presence, and rights of Indigenous peoples, so the dominant view of the future is that of an unoccupied country where future generations of certain humans are denied their very existence and rights – referred to by Krznaric as *temptus nullius*. I do not use 'colonization of the future' lightly, bearing in mind the warnings in Tuck and Yang's influential paper – the epistemological dominance of the powerful today has real, material, destructive consequences for people and nonhumans tomorrow.

Speaking of epistemological dominance, one aspect of the colonization of the future is the colonization of the idea of time itself. The notion of time as linear and sequential (and absolute, as per Newtonian physics) is not exclusively

Western, but in other epistemologies, other conceptions of time are also important. In the context of climate change, Native American (Potawatomi) scholar Kyle Whyte's conception of Kinship Time is especially powerful. In his essay, Time as Kinship (Whyte 2021), Whyte notes that 'when people relate to climate change through linear time, that is, as a ticking clock, they feel peril, and seek ways to stop the worst impacts of climate change immediately. Yet, swift action obscures their responsibilities to others who risk being harmed by the solutions.' Whyte quotes Native American storytellers who tell the story of climate change through changes in kinship relationships, where 'kinship' is not limited to biological relationships but extends to nonhuman beings as well as the landscape through responsibility, that is, 'mutual caretaking and mutual guardianship.' And further, that 'Kinship time, as opposed to linear time, reveals how today's climate change risks are already caused by peoples' not taking responsibility for one another's safety, well-being, and self-determination. Any solutions to climate change will be enacted within a state of affairs that's already rife with irresponsibility. Kinship, as an ethic of shared responsibility, focuses attention on how responsible relationships must first be established or restored for it to be possible to have renewable energy projects that respect Indigenous safety, well-being, and self-determination.'

When we examine a graph such as Figure 4.4, where the time axis is an infinitesimally thin line stretching from past through present to a dire future, we see immediately the urgency of acting upon climate change. This can put us in a state of fear and anxiety, and I speculate that it under the grip of fear, it is easy to give up our agency to demagogues and strongmen. Further, Whyte argues in this essay, when we are trapped in linear time and under the tyranny of the ticking clock, we tend to fall back upon old, status-quo-maintaining ways of doing and being, which have historically been oppressive to Indigenous and other marginalized peoples. To me, Whyte's notion of kinship time is paradigm shifting, located in an epistemology of relationality that sees the world as a complex network of relationships within which we are held and come into being (see also Chapter 6). (This also messes with causality as it is understood in linear time.) Whyte points out that living in kinship time does not preclude acting urgently on climate change. It seems to me that the moment we acknowledge that climate change and the polycrisis are problems of broken relationships among and between humans and nonhumans, we will recognize that any real solution must mend and heal these relationships. What technologies might arise from such a radical re-orientation? Surely such technologies, emerging from such a paradigm, will not oppress or destroy or wreak havoc, as top-down technologies have often done. As I've elaborated in earlier chapters, the onto-epistemological dimension is crucial to both pedagogy and action. Part of the justice aspect of the climate crisis is, after all, epistemic justice. It is interesting to me that Whyte uses the metaphor of the ticking clock – the clock, as elaborated in Chapter 4, is also the metaphor for the Newtonian Paradigm.

So, how might we wrest back the future? How do we push beyond the trap of paradigm blindness to re-imagine radically different futures? The problem of the polycrisis is also a problem of the imagination. How do we free our imaginations to conceive of futures that inspire us to work toward them?

In Chapter 5, I briefly mentioned the power of speculative fiction in this regard. What I describe below is a related exercise that I have done with different groups: environmental justice activists and scholars, an online class on wicked problems in the Indian context, college students at my institution and beyond, and high school students working on sustainability projects.

For this speculative futurism exercise, I lay out a brief framework in which the diagram from Kate Raworth's conception of doughnut economics is helpful: a circle within a circle, where the space in between, the doughnut, represents our workspace. The outer circle is the biophysical threshold informed by planetary boundaries while the inner threshold is the social foundation for human societies, including access to good food, livelihoods, education, rights, etc. The in-between space is where we wish to unleash our imaginations.

Instead of dividing up the imaginative space into sectors such as energy, transportation, education, etc., I provide ten or twelve prompts, each of which is an extremely ordinary sentence, such as 'A man goes to work,' or 'A person is looking out of a window at the city at night,' or 'A person and a nonhuman are in a forest,' or 'A family sits down to a meal.' I divide the class into small groups and ask them to pick a prompt by consensus. Based on all they have learned during the semester, I ask them to take their prompt apart and reimagine every noun and verb. So, for example, one might reimagine 'man' and 'goes to' and the nature of 'work.' Or one might radically rethink the meaning of 'family' or what eating together signifies. Implicit in these sentences are the various sectors such as transportation and energy use, but instead of being chunked up in a reductive fashion, they are entangled in each sentence. The group's charge is to come up with a short story, even a paragraph, that is suggested by the prompt, and inhabits a future in which they would like to live, a future that exists in the space of the doughnut. The only other constraint is that their stories must obey physical law – no teleportation, flying pigs, or ghosts (this is not to deny the possibility of fantasy in this regard, but I have not explored that as yet). This is my attempt toward a diffractive approach to thinking about climate futures and climate solutions.

The results of this exercise vary. In several instances, they only serve to demonstrate the hold that the dominant paradigm has upon our imaginations. So, for example, students might conceive of solar-powered school buses instead of radically re-imagining education. Or they might suggest planting native trees in their neighborhood and stop at that. But in many instances, students and others have come up with far more imaginative possibilities. In one example, participants created a city in which buildings cannot be higher than the tallest and most ancient trees, and where most cars are banned. 'A Man goes to Work' became a cis-het man wearing flowers in his hair, walking to his community solar farm. A group of students re-imagined a city powered entirely by mushroom power, and the cityscape was shaped accordingly. The city as a place for wild animals (of the less dangerous kind) cohabiting in formerly exclusively human spaces came up several times, in one instance the city was arboreal, with human habitations built into trees. In another case, students interpreted a prompt about 'A seasonal community celebration' as a festival to welcome migratory birds back in the springtime. Another group of students imagined an underwater seagrass farm managed by an Indigenous community. One group came up with a conception of a university where there was no student debt, which included a farm managed by students that made the campus self-sufficient. At this reimagined university, learning was focused on environmental education, and professors taught through stories! A group of environmental activists imagined a world where everyone, or almost everyone, is a farmer, whether they dwell in the city or the country. If everyone was able to grow something, what would that do to the market? To the world?

I cannot think of a single instance in which participants did not enjoy this exercise. The activists told me that it was an incredible stress reducer because, for a while, they could live in a better world than the one we inhabit. In these small groups, whether the participants were students or activists or educators, there was shared laughter as well as serious discussion. However, this exercise is still a work in progress. Despite the fact that students can, with help and prompting, use the 5-question framework described earlier to critically and ethically examine proposed climate solutions, this does not suffice for constructing their own real climate solutions. This is apparent not only in the mixed results of the speculative futurism exercise but also in assignments and reflections where students are asked to come up with a real climate solution. The majority of students (81% in one recent class of 16) indicate, through their answers, the *beginnings* of the kind of complex and imaginative thinking skills required for this task. For instance, consider this extract from a freshman's answer:

> My solution to the climate change problem in my town would be to drastically limit the usage of personal vehicles and increase the use of public transportation. My hometown [redacted] has relatively limited public transportation because most people rely on their cars to get around ... Metros and other forms of public transportation should be affordable so that they are accessible to everybody. It would play a role in promoting social justice and equity since everyone, rich or poor, would have access to it, resulting in equality. It would also be assisting individuals who are not well off and do not own cars, to get around the area without making them feel any less than others around, because even people who are privileged would be utilizing the same means. The paradigm employed here is the indigenous people's holistic way of viewing the world. The solution is committed to taking a variety of aspects into account and assisting

everyone and the earth as a whole ... This approach would also be linked to other initiatives being pursued in my area, particularly those aimed at improving greenery ... Other planetary boundaries would be restored, such as species extinction, since by contributing to the climate change issue, they would preserve creatures such as polar bears in the arctic, who are threatened as a result of it.

Affordable and equitable public transportation is, of course, an obvious way to limit carbon emissions, especially if transportation is electrified using renewable energy. Although crucial details are not brought out, such as how this solution demonstrates an Indigenous worldview, or whether 'improving greenery' implies ecosystem restoration, there is an acknowledgment of the need to address multiple issues in an equitable way.

Here is an extract from a different student's answer, who proposes a town in which people live around and inside trees, and where these tree houses are entirely funded by the government and free to all, rich or poor. This sylvan, semi-arboreal existence would allow animals to co-habit with humans.

Using the framework, firstly this would aid in restoration of the carbon cycle because trees will be honored. This would need to be protected and enforced by the government of that state ... The climate cycle would be restored because carbon sinks such as trees and plants would not be cut down, and more trees would be continuously planted for future use. By doing so, more trees will be on the planet which means more of at least one carbon sink. Social equity and justice would be promoted because this would be completely funded by the government ... This solution connects with other solutions such as restoring the carbon cycle and the crisis of homelessness ... This paradigm comes from the view that we were given nature's and earth's gift to work with, and not against. Meaning, the incorporation of nature into our human lives should have naturally been occurring. As with the trees, people will learn to respect, understand, and be knowledgeable about animals because they will not be allowed to be harmed. They would be our neighbors, quite literally.

Here again, there is an acknowledgment of multiple issues being addressed, but the paradigmatic shift is more explicit, which indicates a certain imaginative elasticity. A third student's reflection below is in the direction of education and educational policy change.

The real solution would be to give good curriculum in schools. This includes educating kids in high school too because you shouldn't have to go to college for education in climate change. I believe this will lead to more kids of the next generation finding this subject intriguing and have many people willing to fight against leading companies who don't intend to change their concepts. I believe that not only will college graduates will be able to make good points but highly educated scholars that learned in high school and did research after high school. This solution will reduce the carbon cycle because with more people educated hopefully their actions will make a difference and they'll also take actions in their own life to reduce carbon source emissions. The solution of education should also restore other planetary boundar[ies] because good education will also explain the interconnections with other boundaries and understand that their solutions will have to also not [a]ffect another boundary or positively affect another boundary. I don't believe that my solution promotes social equity if we include business that will lose a lot as a result of changing their beliefs or even smaller companies losing everything. I'm not sure what paradigm my solution is coming from.

Despite this student's uncertainty about the paradigm from which this solution emerges, the above extract also indicates some acknowledgment of interconnections – between individuals and the whole, between education and society, and between climate change and other planetary boundaries.

The reflection exercise from which these extracts are taken is the third of three attempts for freshmen to engage with climate solutions, the first one being the analysis of other proposed solutions through the 5-question framework, and the second being the speculative futurism exercise. The answers indicate the *beginnings* of the kind of complex, ethical thinking that our interrelated social-environmental problems demand, but far more needs to be done to develop more fully the students' imaginative, ethical, and analytical capacities. In future experiments, I plan to engage students early in the semester in examining proposed solutions and helping them discover, as much on their own as possible, the 5-question framework. It would also be far more effective, I suspect, to have at least three runs of the speculative futurism exercise, where, in each iteration, students would use peer and teacher critiques to re-think and/or elaborate upon their proposed solutions. Perhaps, in the last session, we could bring together the ideas of different groups to go from local to global and to traverse sectors and disciplines in order to more fully imagine a different, better world. Making the notion of diffraction and diffractive analysis explicit rather than implicit may also help students become aware that they are being asked to do something radically different that may potentially make a difference in the world.

Engaging with communities in real life through a 'listen-first' collaborative, reflective, and iterative approach may help students co-learn with local peoples how real solutions to complex social-environmental problems might be made manifest in the world. This would be a far richer practice of the diffractive approach. Even the limited engagements with communities and initiatives beyond the classroom, mentioned here and in Chapter 6, already show promise. Given the necessary logistical support, this kind of engagement would likely address more fully the psychosocial action dimension and help enable the cognitive-affective epistemic shift so necessary for transformative learning.

References

Carbon Market Watch. 2022. "Corporate Climate Responsibility Monitor – Carbon Market Watch." 2022. https://carbonmarketwatch.org/campaigns/ccrm/.

Carrington, Damian. 2021. "'A Great Deception': Oil Giants Taken to Task over 'Greenwash' Ads." *The Guardian*, April 19, 2021, sec. Business. https://www.the-guardian.com/business/2021/apr/19/a-great-deception-oil-giants-taken-to-task-over-greenwash-ads.

Dhara, Chirag, and Vandana Singh. 2021. "The Elephant in the Room: Why Trans-formative Education Must Address the Problem of Endless Exponential Economic Growth." In *Curriculum and Learning for Climate Action: Toward an SDG 4.7 Roadmap for Systems Change*, Eds. Radhika Iyengar and Christina Kwauk, 120–43. Leiden: Brill. https://doi.org/10.1163/9789004471818_008.

E.O. Wilson Biodiversity Foundation. n.d. "Half-Earth Project – E.O. Wilson Biodi-versity Foundation." Half-Earth Project. Accessed June 19, 2023. https://www.half-earthproject.org/.

Gadgil, Madhav. 2018. "Sacred Groves: An Ancient Tradition of Nature Conser-vation – Scientific American." *Scientific American*, December 1, 2018. https://www.scientificamerican.com/article/sacred-groves-an-ancient-tradition-of-nature-conservation/.

'Girls' Education | UNICEF.' n.d. Accessed April 21, 2023. https://www.unicef.org/education/girls-education.

Hickel, Jason, Christian Dorninger, Hanspeter Wieland, and Intan Suwandi. 2022. "Im-perialist Appropriation in the World Economy: Drain from the Global South Through Unequal Exchange, 1990–2015." *Global Environmental Change* 73 (March): 102467. https://doi.org/10.1016/j.gloenvcha.2022.102467.

Hoodwinked in the Hothouse. 2021. "Hoodwinked in the Hothouse: Resisting False Solutions to Climate Change," 2021. https://climatefalsesolutions.org/wp-content/uploads/HOODWINKED_ThirdEdition_On-Screen_version.pdf.

International Energy Agency. 2022. "Electric Vehicles – Analysis." Paris. https://www.iea.org/reports/electric-vehicles.

Kothari, Ashish. 2021. "Half-Earth or Whole-Earth? Green or Transformative Recov-ery? Where Are the Voices from the Global South?" *Oryx* 55 (2): 161–62. https://doi.org/10.1017/S0030605321000120.

Krznaric, Roman. 2019. "Why We Need to Reinvent Democracy for the Long-Term." March 18, 2019. https://www.bbc.com/future/article/20190318-can-we-reinvent-democracy-for-the-long-term.

Kübler-Ross, Elisabeth. 2014. *On Death and Dying*. 50th Anniversary Edition. New York, NY: Simon and Schuster. https://www.simonandschuster.com/books/On-Death-and-Dying/Elisabeth-Kubler-Ross/9781476775548.

Lipson, Sarah Ketchen, Sasha Zhou, Sara Abelson, Justin Heinze, Matthew Jirsa, Jasmine Morigney, Akilah Patterson, Meghna Singh, and Daniel Eisenberg. 2022. "Trends in College Student Mental Health and Help-Seeking by Race/Ethnicity:

Findings from the National Healthy Minds Study, 2013–2021." *Journal of Affective Disorders* 306 (June): 138–47. https://doi.org/10.1016/j.jad.2022.03.038.

Mann, Michael. 2020. *The New Climate War.* New York, NY: Public Affairs Books. https://www.hachettebookgroup.com/titles/michael-e-mann/the-new-climate-war/9781541758223/?lens=publicaffairs.

Menton, Mary, Carlos Larrea, Sara Latorre, Joan Martinez-Alier, Mika Peck, Leah Temper, and Mariana Walter. 2020. "Environmental Justice and the SDGs: From Synergies to Gaps and Contradictions." *Sustainability Science* 15 (6): 1621–36. https://doi.org/10.1007/s11625-020-00789-8.

Morrison, Jim. 2020. "An Ancient People with a Modern Climate Plan." *Washington Post*, November 24, 2020. https://www.washingtonpost.com/climate-solutions/2020/11/24/native-americans-climate-change-swinomish/.

OCI Team. 2021. "Report: Indigenous Resistance Against Carbon." *Oil Change International* (blog). August 31, 2021. https://priceofoil.org/2021/08/31/indigenous-resistance-against-carbon/.

Patil, Pallavi Varma, and Melanie Bush. 2022. "'Collaboration across Borders': Comic and Interview." Global Tapestry of Alternatives. July 29, 2022. https://globaltapestryofalternatives.org/newsletters:08:collaboration.

Rangarajan, Mahesh, and Ghazala Shahabuddin. 2006. "Displacement and Relocation from Protected Areas: Towards a Biological and Historical Synthesis." *Conservation and Society* 4 (3): 359–78.

Rights and Resources Initiative. 2018. "New Analysis Reveals That Indigenous Peoples and Local Communities Manage 300,000 Million Metric Tons of Carbon in Their Trees and Soil – Rights + Resources." September 9, 2018. https://rightsandresources.org/carbon-rights-analysis-2018/, https://rightsandresources.org/carbon-rights-analysis-2018/.

Schipper, Lisa, Morgan Scoville-Simonds, Katharine Vincent, and Siri Eriksen. 2021. "Why Avoiding Climate Change 'Maladaptation' Is Vital." Carbon Brief. February 10, 2021. https://www.carbonbrief.org/guest-post-why-avoiding-climate-change-maladaptation-is-vital/.

Schmidt, Jen. 2022. "Clamshells Face the Acid Test." Hakai Magazine. December 2022. https://hakaimagazine.com/news/clamshells-face-the-acid-test/.

Sekhsaria, Pankaj. 2023. "Proposed Infrastructure Project in Great Nicobar Island a Mega Folly." *Frontline*, January 12, 2023. https://frontline.thehindu.com/environment/proposed-infrastructure-project-in-great-nicobar-island-a-mega-folly/article66349362.ece.

Singh, Vandana. 2021. "Imagination, Climate Futures, and the Politics of 'Positivity' | Antariksh Yatra." Antariksh Yatra. August 2021. https://vandanasingh.wordpress.com/2021/08/18/imagination-climate-futures-and-the-politics-of-positivity/.

Swinomish Indian Tribal Community, dir. 2021. *Swinomish Community Visits a Clam Garden.* https://www.youtube.com/watch?v=9dDesE4u07U.

———. n.d. "Swinomish Climate Change Initiative." Swinomish Climate Change Initiative. Accessed June 17, 2023. https://www.swinomish-climate.com.

Temper, Leah, Sofia Avila, Daniela Del Bene, Jennifer Gobby, Nicolas Kosoy, Philippe Le Billon, and Joan Martinez-Alier, et al. 2020. "Movements Shaping Climate Futures: A Systematic Mapping of Protests against Fossil Fuel and Low-Carbon Energy Projects." *Environmental Research Letters* 15 (12): 123004. https://doi.org/10.1088/1748-9326/abc197.

'This Is How Scientists Feel – Is This How You Feel?' n.d. Accessed April 17, 2023. https://www.isthishowyoufeel.com/this-is-how-scientists-feel.html.

Thompson, Tosin. 2021. "Young People's Climate Anxiety Revealed in Landmark Survey." *Nature News*, September 22, 2021. https://www.nature.com/articles/d41586-021-02582-8.

UNEP/FAO. 2021. "Becoming #GenerationRestoration: Ecosystem Restoration for People, Nature and Climate." UNEP – UN Environment Programme. June 3, 2021. http://www.unep.org/resources/ecosystem-restoration-people-nature-climate.

United Nations. n.d. "SDG 12: Responsible Consumption and Production." SDG Indicators. Accessed April 8, 2023. https://unstats.un.org/sdgs/report/2019/goal-12/.

Verlie, Blanche. 2021. *Learning to Live With Climate Change: From Anxiety to Transformation*. Abingdon, Oxon: Routledge.

Watts, Jonathan. 2021. "Climatologist Michael E Mann: 'Good People Fall Victim to Doomism. I Do Too Sometimes.'" *The Observer*, February 27, 2021, sec. Environment. https://www.theguardian.com/environment/2021/feb/27/climatologist-michael-e-mann-doomism-climate-crisis-interview.

Whyte, Kyle Powys. 2021. "Time as Kinship." In *The Cambridge Companion to Environmental Humanities*, Eds. Jeffrey Cohen and Stephanie Foote, 39–55. Cambridge Companions to Literature. Cambridge: Cambridge University Press. https://doi.org/10.1017/9781009039369.005.

Wiedmann, Thomas O., Heinz Schandl, Manfred Lenzen, Daniel Moran, Sangwon Suh, James West, and Keiichiro Kanemoto. 2015. "The Material Footprint of Nations." *Proceedings of the National Academy of Sciences* 112 (20): 6271–76. https://doi.org/10.1073/pnas.1220362110.

Wysham, Daphne. 2012. "The Six Stages of Climate Grief." Other Words. September 3, 2012. https://otherwords.org/the_six_stages_of_climate_grief/.

8 Insights from Other Educators

Reimagining Formal Spaces

8.1 Taking a Critical Look

In Chapter 2, I tried to pose some preliminary answers to the question: 'what is an effective pedagogy of climate change?' In subsequent chapters, I developed various aspects of a pedagogical framework that seems to hold some promise with regard to the need for radical visions of climate education.

But what does 'some promise' mean? How effective is this pedagogy, really? How do I assess it? How applicable is it across disciplines, geographies, cultures, and audiences?

In the last chapter, Chapter 10, I take a critical look at my pedagogical framework and attempt to address these questions. In order to properly engage with these questions, however, I must consider the perspectives of other educators across disciplines, geographies, cultures, and audiences as part of an attempt toward a diffractive analysis.

A downside of working in isolation within a limited context is that the scholar must be careful not to generalize or universalize their experience. What works in one context in a particular spacetime interval may not work in another. Besides, learning through and with other educators and receiving useful criticism is necessary for one's work and vision to develop. Even a single educator trudging alone through uncharted pedagogical territory is indebted to multiple influences and interactions, crucially with students and also with scholarly work, and perhaps family and friends. Such an educator might develop insightful visions from which new and useful concepts emerge, but the work is not solely their own. In analogy with biodiversity, a diversity of influences and interactions (intra-actions?) can make intellectual work – and its practice – richer.

When I first began to work on climate pedagogy beyond 'just the science,' some fellow educators in the sciences reacted with dismissiveness, blank stares, or polite disdain. Others in the humanities and social sciences were more sympathetic, but, with few exceptions, tended not to be very interested in the scientific ideas I was trying to elaborate. Some social science scholars

DOI: 10.4324/9781003294443-8

generously shared their ideas and insights and helped me to understand the sociological and ethical aspects of the climate problem. A couple of people in the sciences were also very supportive and got what I was trying to do. And, of course, everyone in today's academic world is much too busy and overworked. So, the lack of a consistent, diverse, intellectual peer group across disciplines within which to discuss new ideas and from whom useful critiques might be solicited made the going both rough and slow. Meanwhile, the climate problem continued to worsen, as did other major social-environmental issues.

In the course of writing this book, I began to look for other climate educators who were doing interesting things, for two reasons. One was to see whether they might have the time to read a summary of my pedagogical framework and comment on it. The second – most importantly – was to ask them about their own thoughts, experiences, ideas, and challenges with regard to climate education. Through the lenses of these ideas and experiences, I would, perhaps, get a better idea of the usefulness and applicability of my pedagogy in different contexts. But the conversations, I hoped, would be mutually imagination-expanding, positively influencing our work going forward.

This chapter and the next one, therefore, include interviews with educators in different contexts. The last chapter in the book is a reflective-diffractive look at my conceptual framework in the multiple illuminations of influencers and conversations with educators. I begin in this chapter with four conversations with educators in formal spaces.

These are not meant to be a representative sample; I found these educators through my networks and chose them because of their commitment to teaching about climate change, and because they were doing interesting things. They were all incredibly generous with their time and insights, but not all of them had the time to comment at length, and only a few were able to look at a two-page summary of my pedagogy and comment on it. I would have liked to interview many more people in the United States and around the world; however, either my time constraints or those of potential interviewees got in the way – and some did not respond to my request. So, what follows is a small, non-representative sample of educators trying to do brave and thoughtful things about climate education. Because literature on climate education (like most other scholarly work) is heavily dominated by the Global North, my sample is deliberately biased toward educators in the Global South, although I have included interviews with two educators working in North America. Represented in these two chapters, therefore, are educators from Costa Rica, India, the United States, Canada, and Zimbabwe; most are from India, because that is where I have the strongest networks. My description of the work of these educators is based on a preliminary conversation over Zoom, followed by (in most cases) an email Q&A. Quotes have been used with permission.

8.2 A Hope-Infused Pedagogy: Estefania Pihen, Costa Rica, and the United States

Estefanía Pihen owes her perspective on education to a bicultural upbringing in Costa Rica that included holiday visits to national parks and natural reserves. Spending time as a child with children of the Bribri Indigenous tribe near a family farm gave her crucial insights into their value systems and world visions. She also owes much to growing up in a peaceful country with no army. 'In Costa Rica we say that our soldiers are teachers, students, and those that guard nature.'

Estefanía began working as a volunteer in sea turtle conservation at the age of 16, and went on to earn a bachelor's degree in biology, with a specialty in marine conservation. She worked in Las Baulas National Park, and later with CIMAR, a prestigious marine science institute where she studied coral reefs and mangroves. During this time, she witnessed a 'perfect storm for unsustainable touristic development' that dispossessed local people from their lands and culture, and allowed corruption in the construction sector to boom. She saw how rampant and often illegal construction drove deforestation, erosion, coral reef pollution, destruction of sea turtle nesting grounds, and devastation of mangroves. Many local people had sold their property cheaply to developers and found themselves torn from their lands and cultural roots. 'I understood that the lack of proper education generates limited working options for too many Costa Ricans from rural areas, who in turn considered selling their ancestral lands as a viable option for financial stability.'

Her insight, that education is key for maintaining both natural and cultural integrity, inspired her to found a small private K-12 school focused on delivering accessible and often free education to local Costa Rican families. The program of study 'merged high academics with learning about and for social, environmental, and economic local and global sustainability issues.' During 8 years of running the school, Estefanía learned about Education for Sustainable Development (ESD). This has taken her on a journey of learning and exploration that earned her a Masters in social entrepreneurship at the Institute for the Built Environment (IBE) at Colorado State University. During that time, she studied ways to integrate sustainability education in low-income schools in the tropics, co-authoring a toolkit (Pihen Gonzalez et al. 2018) for enabling Green Schools in the Tropics. Centered around the Whole School Sustainability Framework developed at IBE, this place-based, project-based approach employs creative ways to teach and learn using materials easily available in the community, from old car tires to cattle rope. The spirit of this program reminds me of the now famous work of the Hoshangabad Science Teaching Program that was initiated in 1972 to bring science to resource-deprived rural schools in Madhya Pradesh, India.

Isolated examples of outstanding and even radical environmental education exist in many parts of the world. But how to make such education accessible to a wider population, especially among marginalized and climate-vulnerable communities in the Global South? In 2018 Estefanía founded Hahami (https://hahami.org/about/), an organization that supports impactful efforts and projects in transformative education with completed projects in California, Nicaragua, and Guatemala. Creating accredited K-12 curricula, training teachers in effective sustainability pedagogies, and enabling them to design, ideate, and build lessons on sustainability through core subjects are among the services Hahami offers. The website explains the name: Hahami means 'four' in the extinct Chorotega language of Indigenous people who lived in the North Pacific coastal areas of Costa Rica. According to the website, 'The world vision of the Chorotega included a deep respect and a balanced use of nature and its resources.' Hahami refers to the four parts of a holistic approach to learning: Engage, Experience, Educate, and Empower.

Currently, a Ph.D. candidate at the University of California, Santa Barbara, Estefanía, is engaged in research on best approaches and methods for pre-service and in-service teachers on integrating sustainability issues, 'with an emphasis on learning about the current solutions, innovations, efforts, and ideas to tackle and/or mitigate' those issues. She has continued to research the predicaments of teachers in places like her home country. Recent work (Pihen Gonzalez, Estefania and Arya, Diana J. 2021) reported on a study of 42 teachers from Costa Rica, Guatemala, and Nicaragua, participants in a professional development program, who attempted to integrate sustainability into core and non-core subjects. A surprising insight was the degree to which many of these educators, despite not having a background in sustainability education, had integrated such issues into their programs. They were motivated by their lived experience of deforestation, pollution, social inequality, lack of livelihoods, rising violence, and local and national governments that cared little for their citizens. The lack of support from government bodies is a real concern, since partnerships with government programs can provide essential resources and training. 'Such fundamental partnerships might be impossible in countries experiencing political instability, corruption, and regulatory bodies resistant to progressive change in education. This finding raises the question, how can collaboration with such governing institutions be achieved in countries where those in power do not allow it?'

This is a huge question. When climate policy and education policy experts propose plans for implementing necessary changes, are they cognizant of the presence of power hierarchies? Without an explicit acknowledgment of local and global power hierarchies and the injustices that accompany them, is it possible to conceptualize and implement needed change? Governance and education systems in India are also under assault from right-wing, neoliberal forces. It is clear that there are incredibly brave, motivated, and creative people trying to teach about sustainability under extremely difficult circumstances.

Estefanía's pedagogical approach has the following essential aspects:

- Student-centered and solution-based: empowering students as designers of proposals and ideas to mitigate or solve a local or global sustainability issue, she refers to this approach as 'hope-infused,' because students are inspired by case studies and examples of successful efforts. It is also action-based, because the idea behind learning concepts and principles is to apply them to the real world
- Since many educational systems enforce a 'teach-to the-test' mindset, this approach helps teachers integrate ESD into a pre-existing curriculum in innovative ways
- A range of teaching approaches and pedagogics, including 'critical readings and writings; watching and/or creating stimulus activities like documentaries, info graphs, songs or stories; conducting case study analysis; discussing recent critical incidents; conducting simulations; holding debates; developing personal reflections; conducting field work and experiential learning; developing projects; and emulating best practices to reduce individual effects on a natural or human community.' These allow teachers many choices and opportunities to adapt to their contexts
- Fostering sustainability learning principles, including 'critical thinking, envisioning sustainable futures, systemic thinking, empathy and curiosity, co-learning, personal and community values, local and global citizenship, and contextualized learning'

She has applied this approach to teacher training in both formal and after-school programs. In the formal education context, she helps teachers:

- Identify a social, environmental, or economic issue that is relevant to students and their community, or is of a critical global nature
- Find teaching resources about the causes, consequences, and solutions to the issue. Identify how the issue intersects with mandated academic topics
- Design a series of learning activities based on the above resources 'through the application of pedagogies that are conducive to sustainability teaching while also being pertinent for teaching the selected academic topics'

Using this approach, a teacher can simultaneously teach about, for example, fractions, subject-verb decomposition, and local and global efforts to limit greenhouse gases.

In informal settings like after-school programs, the focus is mainly on projects that empower students to become excited about sustainability. Examples include a project in which 'students built furniture with repurposed and waste materials for a recovered outdoor space that was transformed into a garden at a local Boys and Girls Club.' Instead of starting with facts about plastic pollution and the connection with fossil fuels and global warming, students were tasked with imagining pieces of furniture that could be built with specific given materials such as plastic bottles and old doors. It was during the building process that conversations were held to make the connection to global crises apparent.

Teachers who are concerned about climate change and sustainability are hobbled by a number of difficult challenges apart from the problems of local or national power hierarchies. Lack of resources and quality faculty development, lack of support from school administrators and sometimes parents, the narrowed and rigid structure of schooling brought about by standardized testing are some of the challenges that Estefanía has seen teachers face. But there are also successes to celebrate. She names a few: students feeling empowered and hopeful, students with different learning abilities succeeding in academics, projects that are not only learning opportunities for students and teachers but also serve communities' needs, teachers feeling that their profession is meaningful, schools becoming centers for supporting and transforming their communities, and schools with low operational budgets and humble physical infrastructures succeeding at delivering transformative education to their students.

Estefanía encourages fellow teachers to celebrate every success. 'Most people doing transformational work are unsung heroes and they do the work because they know they have to,' she says. 'Making students fall in love with or care for a natural environment, become curious or empathetic of a distant human community suffering the effects of climate change, or excited to share their drawing of their idea for a technical innovation to help with plastic pollution – is extraordinary, meaningful, and worthy of admiration.'

With regard to my pedagogy, Estefanía was intrigued and wanted to know more. But time constraints have forced us to postpone that discussion. I can see some common elements – the need to empower students, the effort to integrate climate and sustainability topics in a transdisciplinary way with core mandated topics, the focus on the local, and on meaningful projects. For me, there are multiple points to ponder, as well as multiple learnings. My pedagogical framework could use a greater connection to the local community, and more attention to experiential learning. My framework may, also, be less useful for the K-12 system where there is even less freedom for teachers to design a curriculum. This is where Estefanía's approach might prove to be a valuable guide, at least until we can have a radical structural reform of education that eliminates the need for teachers to teach climate and sustainability exclusively – and sometimes stealthily – through core subjects.

8.3   Who Are We, in Relation to the World?
Sonali Sathaye, India

'I trained as an anthropologist at Syracuse University, from where I graduated with a PhD in Anthropology in 2002,' says Sonali Sathaye. 'My dissertation investigated mainstream American notions of self and emotion.'

I find it intriguing that she chose, as an anthropologist, to study American culture. Historically, anthropology has been the Global North studying the Global South. But, Sonali says, 'powerful clues to understand what is going on in the Global South lie in the Global North. Increasingly in India, for example, an industrialized arrangement of relationships revolving around the unit of the "individual" is normalized as an inevitability, a foregone rationality; this basis of arranging life sets the context for relationships between humans (in families, schools, friends, romance, etc.) and between non-humans and humans (with forests, trees, rivers, wild animals, mountains).' The Global North sets the agenda, she says 'even in the context of critical thinking – for example, the conversation around climate change and renewable energy, or identity politics, especially to do with questions of gender and sexuality, or psychological politics in terms of the Global Psychology movement.' As I ponder this point, I am reminded of the centuries of European dominance of the planet – so many of our concepts, ideas, approaches, values, that we take for granted, are legacies of colonialism. As Bhambra and Newell remind us, 'our modern world is specifically a *colonial modern* world' (Bhambra and Newell 2022).

Sonali lives in the southern Indian town of Bengaluru/Bangalore. She has spent most of her teaching life in alternative schools in urban areas, or in schools for urban Indians who are typically English-speaking and middle-class. As a PhD student of anthropology, she did field work in two actor-training schools. Her love of theater – she has written and performed plays – informs her work today. What is really interesting to me about Sonali's work is that she does not teach courses focused on climate change – at least, not directly.

> I teach Sociology and theatre to middle- to high-schoolers in and around Bangalore city. Although I have ended up at these schools out of my interest in certain educational approaches – non-exam oriented, open teacher-student relationships, a care for nature and so on – and not because I wish to effect large-scale social change, I have grown to see working with this demographic as a potentially important site in working towards change. This is the population which will go on to have a cultural, economic and environmental impact on their society.

In addition, she says, it is this demographic that models life and consumption patterns on what they see of globalized modernity: on their screens, in shows, in professionalized sporting events, on the news.

Her association with alternative schools has allowed Sonali to design her own courses and obtain funding for courses that transcend disciplinary boundaries. One of these courses also happened to transcend a national border, that between India and Pakistan. The painful partition of British India into India and Pakistan in 1947 – the cost of freedom from British rule – resulted in a vast migration across the border in both directions and the deaths of a million people in the ensuing violence. The impact of that great displacement still reverberates in both countries today.

Sonali's course on the partition of India occurred during the Covid year, 2020, and was held online, with undergraduate students in Ahmedabad, India, and Rawalpindi, Pakistan. The course organizers suggested to Sonali that she focus on the shared culture of the two countries. But it seemed to her that this shared culture was more likely to be 'KFC, Coca Cola and Coke Studio' than the common musical, culinary, and artistic traditions of the subcontinent.

> So, instead, I suggested to the students that we focus on what it is that we actually do share – the natural world; a natural world now under great threat. As part of our seven sessions, I had them describe any relationship they had with the natural world. That led to a rewarding, extremely moving session, where students spoke of sunbirds, streams that are now polluted, a favorite tree, a rock, a shady spot.
>
> Later, students were asked to think about the nature of nationalism: how do human notions of nationhood relate with the natural world, with the land, its ecology, its animals and living beings? If one were to take these into account would "India" and "Pakistan" still be divided in these ways – especially in the context of a phenomenon like the climate crisis which completely overrides human-made boundaries? How might one respond to the climate crisis if these national boundaries were seen for the artificial constructions they are? Earlier in the course I had shown pictures of laid off daily wage workers walking home during the COVID epidemic, photographs that reminded the viewer of nothing so much as photographs of Partition refugees, and asked: If these folks are at "home" (in a discourse in which India and Pakistan are "homes" for separate populations) then where are these people headed to?

There was a palpable sense of togetherness by the end of the course, Sonali says.

Another interesting course that Sonali has designed and taught is on Money. It begins with the question: how is it that human beings are the only ones with any use/need for 'money?' How come it is meaningless to every other being? And, 'If money talks, what does it say?'

> 'Broadly speaking, I was interested in students noticing the degree to which monetary value seems to trump almost every other value in society – from our relation to land and the natural world (the question of "ownership"), to matters we consider personal, such as our relationships with our bodies, our families, our emotions, our talents, to ideas of justice (both in terms of the "equalizing" nature of money, which retains its value regardless of who uses it, but also in the ways that the profit motive moves in this arena). What does it mean to "own" something?

Do we recognize any other means of "ownership" or does money trump all of those other ways – e.g. do the Dongria Kondh [an Indigenous tribe in eastern India (Orissa)] have any "rights" over a forest or a mountain that has been their home, that they have lived happily on and with for thousands of years, or does a corporation's monetary "ownership" legality upend those rights?'

The course was arranged so that every week students explored the topic with Sonali and one teacher of either Economics, Biology, Philosophy, or History/Law. This led to interesting conjunctions: when exploring the question of money in the natural world, for example, she took her teenage students on a walk through the nearby Bannerghatta National Park, where they noticed that the forest generated no waste. Anything discarded by one species was used by another. She posed the question: 'Is it also merely coincidence that money is also completely absent in the natural world?' And, 'What, if any, is the connection between our mountains of waste and our mountains of money?'

Sonali is currently teaching a course about food to high school students. She is interested in exploring with them the extent to which food choices are dictated by the market. The India I grew up in, during the 1970s and 1980s, had almost no plastic packaging or junk food – the economic liberalization of the 1990s opened the floodgates to international junk foods and vast mountains of waste within and outside cities. There are more obese people now in India than ever before. Sonali wants her students to engage with questions around hunger as well. Why is it that vast numbers of people in the world are malnourished? What is the global food system all about?

Sonali is working with these high schoolers and another teacher on a project with middle-schoolers in their local government school. Government-run schools are typically attended by students from the lower rungs of the economy. 'Passing the growing piles of rubbish outside the school building on my way to work every week, I had thought to engage the children of the local school on notions of aesthetics in their environs, and food and waste.' But the principal wanted the children to be taught English.

Consequently, we have been trying to teach English through conversation, drama, little walks and other activities around the topic of food, waste, plastic .... Both courses have involved gardening/growing food, apart from a project, such as writing a paper or a graphic novel. Also, both involve ethnography – students are encouraged to speak with the wider community in which they find themselves (on the outskirts of Bangalore city, in a semi-rural area). Finally, some theatrical work forms part of the activities of each course.

These multiple modes of engagement help students see taken-for-granted concepts afresh. But this is a difficult task. I ask Sonali why this is so.

'Students seem to perceive the world through an increasingly subjectiv-ized lens,' she says. 'Things are viewed at the level of the individual and their exigencies, contingencies, their helplessness in the face of circumstance.'

Information about, for example, the chicanery of fast food corporations, doesn't evoke outrage. Reactions she has seen range from a 'mental shrug' that seems to imply a tacit acceptance of these issues, to sympathy for employees of the corporations, who would be out of a job if the corporation were to cease existing. She gives the example of a student in her Money course, who was uncomfortable comparing the relative value of the work of a farmer versus a phone game designer with regard to the health of society and the planet. The student didn't want to judge any kind of job, 'because people had to work for their families.'

> The logic of these responses is based on seeing actions through the lens of the individual, not at a systemic level. Typically, students seem to accept the existence of the profit-making Corporation as an organizing agent of society; rarely do they appear to question the reasons for such organiza-tion, or to envisage a life for themselves in which they buck the system – even in such relatively trivial matters as shopping for "fast fashion" or eating fast food.

Sonali feels that this is primarily because students are not taught to think sys-temically, let alone recognize how human-natural systems interact. 'This is in keeping with our global and national approach, which appoints one commis-sion to look into air pollution, another to look into policies around "conflict" zones between animals and humans, a third to look into the effects of eating a processed diet and so on. We seem to be unable to see the whole as one organism – the whole of society, the world, the planet.'

She feels that 'young people's experience of life is shrinking.' The tragedy is that they will not miss what they have never known. This 'extinction of experience,' to evoke Robert Pyle, and the fractured, diminished worldview it engenders, is what motivates her to create the kinds of courses and ex-periential learning opportunities that she does. Climate change comes into the picture as part of a deeper, larger problem. She is interested in what lies beneath the proliferation of global social-environmental issues. To use the language of my pedagogical framework, she is working in the epistemological dimension, helping students see that our conceptualizations of reality come from paradigms. The fragmentation of reality, including the separation from other species and the inability to see systems at work in the world, is, to me, symptomatic of the Newtonian, mechanistic paradigm. Globalization and the enduring coloniality of power and knowledge have made this the dominant paradigm of modern industrial civilization.

In this sense, 'indirect climate education' has value because it reveals the climate problem as a symptom of something deeper – a predatory socio-economic system that inculcates alienation of humans from the rest of Nature. Sonali says that she is not interested in teaching climate change directly to students for this reason, but there's more. 'We have chosen to live with symbols rather than actuality ... so we choose money for its own sake, rather than for the things it is supposed to be able to facilitate. We choose experiences in order to represent them on our social media; our leaders come to power because of the names they give themselves; we eat "food-like" substances rather than actual food.'

Sonali's approach makes apparent the hitherto invisible constructs of Modern Industrial Civilization's 'consensus reality.' What we take for granted as real, grounded, set-in-stone, is, in fact, ephemeral, shifting, and divorced in crucial ways from what matters in the world.

But Sonali's experience is also potentially valuable to teachers in more conventional formal settings, who wish to introduce climate change in the context of their disciplines. Even when they don't have the freedom to design their own courses, how can they harness their creativity within their constraints to engage the students through the epistemological dimension? After all, what we want students to experience is the epistemic shift referred to in transformative education theory. My own experiments with 'physics theater' in my classrooms indicate that drama is a powerful learning tool, and that experiential learning is deep learning. We need more experiments, more research, before we can come to any grand conclusions. Sonali's work points to the need to depart from conventional classroom settings to deeper and more creative engagement with students on the existential crises we face. Difficult though it is to free students from the trap of the dominant paradigm, the effort is more than worthwhile. The non-traditional approaches that Sonali describes help enable self-searching, good debate, and the possibility of relating differently to the world. 'Teaching is itself a joy – the laughter, the art, the listening and being guided by young people on how to teach them the things they should know,' says Sonali.

How do students respond to Sonali's classes? Students have come back to her at the end of the semester, and sometimes a year after the class is over, to tell her that they can now make more sense of their experience of the world; that they can see where they stand within socially constructed systems like patriarchy and gender, for example. Her sense is that she has at the very least helped them start questioning the taken-for-granted, 'natural,' default conceptions that shore up modern civilization, such as money, nature, and what constitutes food. This is the first and essential step toward reframing their relationship to the world.

It is this last point – how students relate to the world – that Sonali asks about after reading the summary of my pedagogy. Have I explored, she asks, how the individual-structural relationship plays out? Do my classes discuss the ways in which we are implicated, absorbed into the current state of affairs? How do we understand who 'we' are in relation to the world, natural and social? At one level, this is about our choices in terms of food, transportation, major, job ambitions, etc. But it goes deeper than that. If the individual and society co-construct each other, in part through the dominant paradigm interacting with a person's innate traits, along with the influence of familial and affinity groups, and if this co-construction of the individual leaves out other groups of humans, other species, and the web of interconnections that is life on Earth, then, to ask 'who am I in relation to the world' is to question one's identity and being in a very fundamental way. And this can be the start of an epistemic shift within the student and the community of the classroom.

As I reflect on my work in response to Sonali's question, it occurs to me that I can, perhaps, find more explicit ways for us to explore 'who we are in relation to the world, natural and social.' Many of my efforts are in this direction, at the intersection of the epistemological and psycho-social dimensions. These include making a concept map at the start of the semester to introduce ourselves to each other, showing us in webs of relationships; making space for students to share how they feel about what they are learning; building a relationship with each student so that they feel heard and supported; doing empowerment exercises in the classroom so that students see that their ideas and opinions matter; helping students apply the concepts we learn to question the socio-economic system; having students work on projects in groups, especially projects that have a community service aspect; having students teach each other and the class; doing self-reflections and other exercises that inculcate metacognitive skills.

And yet, these only indirectly address the question: given all that is troubled in the world, how, then, should we live?

8.4   Pedagogy, Policy, and Paradigm: Yovita Gwekwerere, Canada, and Zimbabwe

As a biology lecturer in Zimbabwe, Yovita Gwekwerere first came across journal articles on climate change in the mid-1990s. Being a new teacher, she was experimenting with various teaching approaches, and students enjoyed reading and discussing such journal articles. But outside her classes, the subject was off the radar of most people. When she went to the United States for her Ph.D. studies, she was surprised to find a similar silence around the subject of climate change. In the last year of her doctorate, she worked on a project exploring how ecology was taught in schools, and this reawakened her interest in climate change. Later, during her postdoctoral work, she began to attend events organized by Greenpeace, and this gave her heart that there were people who cared about the crisis.

Yovita now teaches and conducts research at Laurentian University in Canada, where she is an associate professor in the School of Education and the sole expert on science education. For the past 15 years, she has been integrating Environmental Sustainability Education into science pedagogy in all her courses and conducting research on climate education, including the understanding of Environmental Education among preservice teachers. Recognizing the crucial role of policy in effecting systemic change, she has contributed to the state of Ontario's development of an Environmental Education policy, and has introduced it to preservice teachers. She has conducted projects with colleagues where students integrated the environment into Science, Music, and Literacy, and has enabled B.Ed. students to launch an Outdoor and Environmental Education club.

In 2014, I started collaborating with other like-minded educators in Ontario, and we have been working together to advocate for the integration of Environmental and Sustainability Education in the Teacher Education programs across the country. We collaborate on research, and we have co-edited a book on Environmental Sustainability Education in Teacher Education in Canada. Starting in 2015, I have been collaborating with colleagues in Southern Africa to integrate environmental sustainability and Climate change education across the school curriculum.

It is a peculiar blindness of the Western education system, globalized through colonialism and market forces, that education about the environment is not already a requirement for teacher training. More often than not, environmental programs are often among the first to be cut when there is a financial crisis, as Yovita herself has experienced. This echoes what researcher Christina Kwauk has said about eco-literacy being considered less important than literacy (Kwauk, 2020) as though the two are mutually exclusive. To live in a world in which we are indebted to multiple lifeforms and natural cycles, and to omit these essentials from the education of teachers and children seems to be a monumental oversight. During our Zoom conversation, Yovita and I shake our heads over this.

Yovita believes that the current environmental crisis is 'too urgent to leave education to formal schooling. Everyone must be educated.' When a School of Environment opened at her university in 2014, she had a cross-appointment there, until 2021 when the School closed. During her time there, she developed an environmental education course for education majors and non-majors.

'This course was designed for teaching Environmental sustainability in the formal and non-formal settings and I called it "Environmental sustainability education for the masses,' she says. 'The course focused on developing environment mentors where students were required to teach their peers and families for one of the class projects. I also taught an introduction to Environmental Studies course, a first-year elective course that was open for students from across programs. I am still teaching this course that is very popular with students from different programs.'

Yovita's advocacy efforts with her colleagues to require Environmental Sustainability Education in teacher education programs in Canada has paid off. 'In 2022, the Deans of education across the country published an Accord on Education for Sustainable Futures in which they acknowledge the need for education, and specifically Teacher Education programs to address the environmental and climate crisis in the curriculum. The accord is critical in that it informs policy, making it mandatory for programs to address the issues.'

Even in a country like Canada, where educators have more resources and perhaps more academic freedom and the power to affect policy than in countries of the Global South – despite a current government bent on neoliberal economics and fossil fuel development – there are challenges. Yovita explains that environmental and climate literacy are integrated in the school curricula of most Canadian provinces. School-age children are therefore more environmentally literate than most adults.

> However, challenges experienced in schools include funding, professional development and time constraints. The situation is different in Teacher Training programs where environmental/climate literacy policy is non-existent, hence it is up to individual instructors to develop courses or integrate the environmental content in their courses. Since individuals at university level are specialists in certain fields, it is challenging to find enough individuals who have specialization in the areas of environmental/climate education.

Perhaps the new Accord on Education will change things. This aspect of Yovita's work points to the potential of teacher education scholars at universities to effect policy change at the state and national levels, at least in countries where this is politically feasible.

In 2015, Yovita began collaborating with colleagues in Southern African countries on integrating environmental sustainability across the curriculum. I am especially interested in a paper (a book chapter) she published in 2021 with Overson Shumba, professor of education at Copperbelt University in Zambia (Gwekwerere and Shumba 2021). The paper analyzes roadblocks to transformative education for climate action in southern African countries, and identifies the following:

- A historical antecedent that has impinged on the framing and internalization of ESD-inspired curriculum reforms
- Lack of resources to support teacher agency and leadership in transformative learning for climate action and
- Lack of curriculum connections to local traditions and real-life challenges and vulnerabilities confronting the learners

A 2011 paper by Overson Shumba (Shumba, Overson 2011) makes clear the historical antecedent:

> Western education in Africa has historically been part of a process of enclosure, a consequence of colonisation. As a result, Western education

promoted initially through missionary activity has served by and large to supplant local traditions viewed as primitive, tribal, and backward. Post-colonial education systems of education have not shaken off the messianic message of Western education and continue to face challenges related to relevance.

In their 2021 paper, Yovita and Overson Shumba suggest a pedagogy inspired by the Afrocentric concept of Ubuntu. Her colleagues in Southern Africa have been working with this philosophy to think about transformative learning and addressing Sustainable Development Goal 4 (inclusive quality education and lifelong learning).

What is Ubuntu, and why consider it as a foundation for climate and environmental education? The word is most often translated as 'I am because we are,' a sentence that signifies relationships and connection. South African scholar Michael Onyebuchi Eze explains it thus in his book *Intellectual History in Contemporary South Africa*.

This idealism suggests to us that humanity is not embedded in my person solely as an individual; my humanity is co-substantively bestowed upon the other and me. Humanity is a quality we owe to each other. We create each other and need to sustain this otherness creation. And if we belong to each other, we participate in our creations: we are because you are, and since you are, definitely I am. The "I am" is not a rigid subject, but a dynamic self-constitution dependent on this otherness creation of relation and distance.

This concept stands in stark contrast to the lonely individualism of the famous Cartesian dictum 'I think, therefore I am.'

Yovita's contention is that:

In the face of the climate crisis, it is clear that we cannot continue to look for answers from the Western education philosophy that has led to the current crisis. I believe that we can seek wisdom from other traditional non-Western philosophies from Indigenous cultures to come up with solutions to address the current environmental and climate crisis. The Ubuntu African philosophy is just one of many other Indigenous ways of knowing that do not emphasize individualism (something that is perpetuating the current environmental and climate crisis). I am currently investigating ways to use Ubuntu inspired pedagogies with secondary students in Southern Africa.

Reflecting on my conversation with Yovita later, I see her work on Ubuntu-inspired approaches to climate and environmental education as particularly relevant to the epistemological dimension that is one of the four pillars of my pedagogy. I am also inspired by her advocacy, as a scholar of education, for change in educational policy to facilitate meaningful climate education. There is much that education departments at universities can do to foster the structural changes so urgently needed. In the classroom, I have sometimes used a concept-mapping technique for students to understand how they are connected, through family and friends, schools, towns, and interest groups, to a wider range of people than they are generally aware of. As an exercise in mapping our ripple effect, this helps students to go from the isolated individual to an agent of collective change. Perhaps this exercise needs to be done with teachers and scholars as much as with our students.

Yovita's response to my pedagogy is encouraging. 'We need more scientists to be collaborating with educators in efforts to promote environmental/ climate literacy,' she says. And, in fact, this is a lack I see in much of the scholarly work from educators. Climate science is developing fast, as are multiple new climate and energy-related disciplines across the social sciences and humanities. New knowledge and new insights must inform climate education. We must think about how to fill this gap. Yovita also appreciates the transdisciplinarity of my approach. She encourages me to team up with an education scholar to further develop my pedagogical framework. This is an important next step for me.

8.5   A Crucial Role for Teachers' Unions: Karen Trine, Chicago, US

When I spoke with Karen Trine on a recent Monday afternoon over Zoom, she was in her classroom, talking with student members of the school's Green Team. Karen is a chemistry teacher at a selective enrollment high school in Chicago. I first met her, along with her other colleagues from the school, at an International Education conference at the University of Chicago in 2019, which was themed around climate change. Following a wonderfully rich discussion there, Karen and I have recently re-started our conversation on climate education in US high schools. Although Karen wasn't able to find more time to talk or respond over email at length, our short conversation for this chapter was enlightening and valuable, opening doors to various possibilities.

Karen began her journey as a young educator teaching chemistry, physics, and mathematics in the Peace Corps. She taught for two years in rural Burkina Faso at the junior high and high school level. She also acquired some cross-disciplinary experience, teaching an English-plus-Environmental Science class for two years at a teachers' college in rural Sichuan, China. Back in Chicago, she has been teaching at selective enrollment high schools for 16 years.

As an educator and mother of young children, Karen is deeply concerned about the climate-changed future. She tries to bring climate topics into her chemistry classes whenever possible, by using climate examples as context for chemistry topics, or as extra credit special projects. She feels there is never enough time or resources to do justice to the subject, and that this piecemeal way to introduce the problem is not sufficient. What would work better, she thinks, is a required class on climate change that is taught in a multi-disciplinary way. We chat for a while about how that might come about. It is not a secret that major infrastructure changes are needed to bring climate change usefully into the US curriculum. But how to accomplish these? There are the complexities of region-by-region difference in curricula, which, by themselves, are not a bad thing – education should have relevance to place. But among several states and districts in the United States, climate denialism is widespread. And even in places where this political barrier is absent or weak, there are other structural problems, including lack of access to adequate teacher training. Karen acknowledges the prevalence of climate topics in the Next Generation Science Standards (although not all states have adopted these) but there are too many institutional and other barriers to their proper implementation.

During our conversation, I recall a climate education bill that had been proposed in the twenty-teens by Senator Ed Markey. Given Trump's rise to power, the bill at the time had no hope of becoming reality. If passed, the bill would have provided, among other things, funding for professional development of teachers. The early form of the bill had its drawbacks, for instance a narrow, technocratic focus on climate science and carbon emissions without an acknowledgment of the polycrisis and the underlying justice dimension – and it also lacked awareness of the need for climate education research. Karen and I spoke briefly about what enlightened federal legislation might do today. At the very least, it would help bridge the gap between need and preparation for US teachers. It might also mandate a required interdisciplinary course in high schools. I offer a thought: that perhaps we need *both* a required interdisciplinary course and a climate-across-the-curriculum approach so that the climate problem is not isolated from other disciplines. But what about middle school and elementary school? Should there be a mandatory climate change course in those grades as well?

Karen thinks that we may have to be a little careful about the early grades. For young children, being exposed to the full horror of the climate crisis before they are emotionally prepared for it might just push them into hopelessness, despair, and apathy. What, then? Karen suggests that nature education – specifically, through immersive experiences in Nature – needs to be a requirement in early grades. 'Kids first need to identify with nature as younger children, to have the intrinsic motivation to work hard to improve currently projected climate outcomes,' she says. This makes sense to me. We can only protect what we love, and fostering the connection to our more-than-human world is likely to help engender that motivation. Karen tells me that her family has a summer business – managing a small, sustainable fishery in Alaska, which involves being out at sea or in the coastal wilderness for extended periods of time. It has been transformative for her young children. She wishes all her students had access to such experiences.

Meanwhile, what works in the classroom and what are some of the challenges?

> What works are projects that give kids hope. They seem to know how bad it is already and wonder why all the adults don't. Challenging aspects include not having access to system-wide solutions, but only personally relevant actions like energy conservation or recycling, etc. Kids understand that this is a political issue, but don't have a lot of political power.

It's not just the kids, of course. Many of us recognize that we need system change, and yet we are embedded in systems that are deeply entrenched and seem impossible to shift. 'I feel stuck not knowing what I can do to help a lot of the time. In my role as a teacher, I tend to console myself with the idea that by helping kids understand the importance of climate crisis issues, I'm helping in the long term … but in the short term – the whole problem feels very big and very overwhelming.'

This is, of course, one of the reasons why collective action is so important. Karen sees a crucial role for teachers' unions. She feels that unions in general are starting to recognize that climate change is a union issue (which is echoed in this news article https://www.theguardian.com/money/2021/sep/20/labor-climate-allies-green-union-jobs). She herself is active in the Chicago Teachers' Union, and tells me of an important Green Schools Campaign they have launched. Far from being climate-ready, Karen tells me, many Chicago schools have failing infrastructure. The campaign's website states three aims:

- Address the Climate Crisis and Environmental Racism
- Improve Our School Buildings
- Green Careers to Our Communities

From their website:

> Our schools are dirty, outdated, and full of harmful toxins. Too often, schools on the South and West sides with the most renovation needs are not prioritized. Who is first in line is often based on political opportunity and what looks good in an election year. This type of planning is racist. We need a process that is inclusive, just, transparent, and prioritizes Black and Brown communities and brings those schools into the 21st Century to avoid the worst impacts of the global climate crisis and carry us into a sustainable future.

The criteria for green schools as laid out on their website include not only climate-ready renovations, but also an 'empowering curriculum' and 'green spaces and community gardens.'

Later, I reflect on my conversation with Karen. We both appreciated the chance to talk about the challenge of teaching climate change. How often do concerned educators have the opportunity to share how they feel, let alone brainstorm together? The psychosocial dimension is applicable to teachers as much as to students. Karen feels that the grief and depression that result from learning about climate change need to be addressed in the classroom. For that to work, teachers have to deal with these too, and in my experience, that can best be done collectively, through networks and collaborations.

My conversation with Karen helps me get a sense of how difficult it is at the K-12 level to teach climate change. There is growing research documenting both challenges and successes in teaching climate change in the United States (Plutzer et al. 2016; Henderson, Joseph and Drewes, Andrea 2020; Busch 2021). While Karen is lucky not to have this problem at her school, a fixed curriculum and pressure to teach to the test are among the larger systemic barriers in many school districts. It seems clear that my pedagogical framework is unlikely to fit into a traditional, siloed education system with compartmentalized disciplines. In schools where there is some flexibility, I am hopeful that some aspects of it – the use of stories, and the repeated use of visual tools such as Figure 3.1, that can help make piecemeal climate topics in a traditional course feel connected, for example – can inform courses across the spectrum from the natural to the social sciences and humanities. For instance, in a Chemistry course, the use of a climate story can bring out all the four dimensions of the climate crisis, and even if the class is constrained to explore only the chemistry-relevant portion, the students are at least made aware that there are other, equally important aspects, from justice concerns to socio-economic issues. Karen's suggestion of teacher-ready materials for such adaptations is very much to the point. It would be particularly interesting to see if my pedagogy could be of use for a transdisciplinary mandatory course, if such a course is piloted in US schools. But all these innovations require further research and the political will to enact system change.

More than anything, our conversation energizes us both. In a later email, Karen tells me that after our conversation she has been thinking deeply about the possibility of an interdisciplinary, co-taught course on climate change at her school. For my part, I go find my local educator's union and discover a climate action group, which I then join. Just from an hour's Zoom conversation and a few emails, we both feel inspired, hopeful and ready to take some action.

References

Bhambra, Gurminder K., and Peter Newell. 2022. "More than a Metaphor: 'Climate Colonialism' in Perspective." *Global Social Challenges Journal* 1 (aop): 1–9. https://doi.org/10.1332/EIEM6688.

Busch, K.C. 2021. "Textbooks of Doubt, Tested: The Effect of a Denialist Framing on Adolescents' Certainty about Climate Change." *Environmental Education Research* 27 (11): 1574–98. https://doi.org/10.1080/13504622.2021.1960954.

Gwekwerere, Yovita N., and Overson Shumba. 2021. *A Call for Transformative Learning in Southern Africa: Using Ubuntu Pedagogy to Inspire Sustainability Thinking and Climate Action*. Leiden: Brill. https://doi.org/10.1163/9789004471818_011.

Henderson, Joseph, and Andrea Drewes, eds. 2020. *Teaching Climate Change in the United States*. Abingdon, Oxon: Routledge. https://www.routledge.com/Teaching-Climate-Change-in-the-United-States/Henderson-Drewes/p/book/9780367179472.

Kwauk, Christina. 2020. "Roadblocks to Quality Education in a Time of Climate Change." *Brookings* (blog). February 25, 2020. https://www.brookings.edu/research/roadblocks-to-quality-education-in-a-time-of-climate-change/.

Pihen Gonzalez, Estefania, and Diana J. Arya 2021. "Understanding ESD: Perceptions and Views from Guatemalan, Nicaraguan, and Costa Rican Educators." In *Environmental Sciences Proceedings*. Vol. 15. Environmental Science Proceedings. https://doi.org/10.3390/environsciproc2022015045.

Pihen Gonzalez, Estefania, Stephanie K. Barr, Brian H. Dunbar, and Jennifer L. Schill 2018. "Green Schools in the Tropics: A Toolkit for Schools on a Budget." Fort Collins, Colorado: Institute for the Built Environment, Colorado State University. https://ibe.colostate.edu/wp-content/uploads/2019/04/GreenSchoolsInTheTropics2019edit.pdf.

Plutzer, Eric, Mark McCaffrey, A. Lee Hannah, Joshua Rosenau, Minda Berbeco, and Ann H. Reid. 2016. "Climate Confusion among U.S. Teachers." *Science* 351 (6274): 664–65. https://doi.org/10.1126/science.aab3907.

Shumba, Overson. 2011. "Commons Thinking, Ecological Intelligence and the Ethical and Moral Framework of Ubuntu: An Imperative for Sustainable Development." *Journal of Media and Communication Studies* 3 (3): 84–96.

9 Insights from Other Educators

Climate Education Outside the Walls

9.1 Beyond the Classroom

My four interviewees in the last chapter work in different kinds of formal spaces, and yet in each case they are reaching out beyond the four walls of the classroom. Estefania's philosophy of engaging students with the local communities in school and after school while paying attention to teachers' needs, Sonali teaching her students to literally learn from the world outside the classroom and consider their relationship to it, Yovita's focus on teacher training, not just within the university space but also beyond it to national education policy, Karen's insights on the challenges within the high school classroom, including failing physical infrastructure and the role of teachers unions in engaging with climate change – all of these indicate that meaningful climate education must break down the walls between educational institutions and the world at large.

An additional imperative, is, of course, that it isn't just students in K-12 and college who need meaningful climate education. So do teachers, policymakers, government and corporate leaders, and, especially, communities that suffer the most from climatic impacts. In this chapter, we meet three educators whose work reaches further beyond formal spaces.

9.2 A Practice of 'Humble Wonder' – Yuvan Aves, Chennai, India

Yuvan Aves is a naturalist, nature educator, writer, and environmental defender in the coastal city of Chennai, India. He spoke to me over Zoom in March 2023, about his work with children and adults in the area of nature education, which includes, but is not limited to climate change. He uses the term 'Nature' advisedly, he says; his sense of 'Nature' includes human beings and is not intended to reinforce the problematic human/Nature binary.

Despite his youth, Yuvan has experienced and overcome a number of challenges that have helped him acquire a drive and a maturity that is rare even among much older people. He grew up in Chennai, where the forested

DOI: 10.4324/9781003294443-9

environs of an alternative school run by the Krishnamurti Foundation nurtured his interest in the wild, an interest his mother encouraged. At the age of 16, he ran away from home to escape an abusive stepfather and took shelter with a teacher who had been one of his early inspirations. The teacher had opened a small school in a village, and this is where the boy Yuvan started on his first steps toward becoming a nature and climate educator. Traumatized and unable to sleep, he would wander the fields with his binoculars, provoking the curiosity and interest of the village children. Soon, he was taking them on birdwatching trips. Later, he went on to finish school and get a college degree, but his path turned out to be quite unconventional. Today, he runs the Palluyir Trust for Nature Education and Research, a unique organization that aims to mainstream place-based, Nature-based, and outdoor education throughout the school system and beyond. What is most interesting about Yuvan's work is its range: he and his team work with children from all economic back-grounds, as well as with teenagers and young adults (the age range is 2.5 years through college). A key focus is marginalized communities, including domestic workers and the fisher communities on Chennai's coast. The idea is to not just generate awareness but also nurture a sense of belonging and a drive to advocacy. Training future trainers is a crucial part of this model. Many former students are now nature educators.

Yuvan has clearly thought a great deal about pedagogy. He tells me that the three foundational aspects of his approach are:

Direct participation – nature walks and observations followed by recording, sharing, art, advocacy, and projects that include the student becoming the teacher for their community or circle
Local Relevance – starting where you are, and understanding the landscape, the people and nonhumans who call it home. All the work he does with fisher children, or internships like the Urban Wilderness Walk, is situated in the ecological-social context of Chennai
Interconnectedness – relating the human to the nonhuman, and the global climate problem to what is happening in the here-and-now

Take, for example, one of their activities, in their Climate, Biodiversity and People Curriculum. This is about migratory birds and climate change, and, like other material Yuvan and his colleagues have created, is free to down-load from their website. It includes learning about and making observations of migratory birds in the locality, and understanding how they are affected by climate change. But it doesn't stop there. Part of the learning is also to communicate and spread the information – each participant takes a group from their community on a migratory bird walk, transitioning from student to teacher.

Yuvan shows me a website, inaturalist.org, that he has used as a Citizen Science tool for developing a project called Biodiversity of Chennai. 'It is difficult to love something you cannot name,' he says, clicking to the project site, where, with the help of students and Nature educators, he is building a database of observations of fungi, insects, plants, birds, and animals. He started this project with five people, and this year there are more than a 1,000 working on it. He points out that such a database can also be used for advocacy; for example, if the government is attempting to push through the environmental clearance on a project, any claim that biodiversity is absent can be contested.

The sociological dimension is not neglected in this pedagogy – in fact, the issues of inequality and marginalization are central. One of the activities in the Climate, Biodiversity and People Curriculum is to learn about societal privilege. Based on work by Sylvia Duckworth (Center for Teaching, Learning and Mentoring 2022) and adapted for the Indian context, this activity allows the learner to recognize their privilege and explore how privilege is related to climate change. Thus, climate change can be seen as fundamentally related to power and inequality. But here, also, learning must translate into meaningful action. Through a mind-mapping activity, learners examine how they can use their privilege to flatten hierarchies in their world.

Palluyir Trust offers a Youth Climate Internship; Yuvan also offers an Urban Wilderness internship for college students via the Madras Naturalists Society. Graduates who go on to become Nature Educators include people from marginalized communities.

I ask for details on working with marginalized groups. He tells me about making the beach a learning space for fisher children. The classroom is the sand, sea, and sky. There is garbage too, but also life. The hierarchy of the traditional classroom is absent. Spatially as well as pedagogically the beach allows for a 'multidirectionality of engagement, where you and I might be speaking to each other and to the heron over there,' unlike a traditional classroom, where everyone is looking in one direction. Here's a crab, there a bird, and all of them are teachers. According to Yuvan, Nature is a great equalizing force. As the children build a deep sense of belonging with the environs they call home, two things happen. One, their relationship with other species and the landscape tends to dilute their sense of social oppression. Two, connection is a strong intrinsic motivation for learning. Yuvan refers to Robert Pyle's notion of 'extinction of experience,' the loss of human-Nature connection (Pyle, Robert 2011). Children who have gone through this kind of learning often demonstrate an astonishing increase in literacy and writing skills, which gives them confidence and better prospects. In fact, this has been such an inspiring outcome that Yuvan is currently working on a proposal for the state government for a research study in seven government schools, initiating an 8-month Nature and climate program, to see how it affects intrinsic motivation to learn, along with attendance, academic performance in science and language, mental health, and conservation attitude. If they can establish these, this program may well go statewide.

This last point reminds me of Christina Kwauk's identification of a major roadblock in climate education (Kwauk 2020): that eco-literacy is not valued by mainstream education, which focuses exclusively on literacy. But, as Yuvan's work indicates, literacy can be dramatically enhanced through place-based, Nature-based education. The two are not mutually exclusive.

One of the things Yuvan has learned from climate-vulnerable communities like the fisherfolk, or the domestic labor and Dalit communities is that structural oppression doesn't have to be taught – these communities are well aware of this central, daily aspect of their lives – and, at the same time, he has found that the people he has trained have a strong drive to become agents of justice. This is in stark contrast to the teens of the upper middle class and rich who go to expensive schools, many of whom seem to be afflicted by what Yuvan refers to as a 'bulletproof individualism.' They are not concerned about birds or turtles, but mostly about making money. However, rich or poor, he speaks highly of the capacity of children to be enchanted by nonhuman Nature. 'I've not met a single child under 7, under 10, who is not enthralled, spontaneously engaged with the more than human, the real world.'

And yet, there are important nuances to working with marginalized groups. Yuvan tells me about a spectacular biodiversity hotspot, an estuary about 100 km from Chennai, where two fishing harbors are proposed. The estuary is a nesting ground for endangered turtles, busy with birds of many species, with seagrass meadows, mangroves, and an oyster reef. Forty kilometers out into the ocean, sperm whales congregate to calve. Two of the local fisher communities are in favor of the proposed fishing harbor, although this would pose a serious threat to the ecosystem. It turns out that people in these communities now own fishing trawlers. 'Capitalism gets them,' Yuvan says. 'It is problematic to have a homogenous idea of "Indigenous."' He tries to see their perspective. 'Perhaps if I was a trawler owner I would want a harbor.' A trawler, though expensive at Rs. 8 million, is a doorway to get rich. Recently, a friend of his who had been doing turtle counts in the estuary was forbidden to do so by the fisher communities. But, Yuvan tells me, there are people opposed to the project. Further inland, the communities who depend on the wetlands have a vested interest in the health of the ecosystem. Hundreds of women from these inland communities come to harvest oysters, crabs, and lugworms. But they are afraid to speak up. 'There are inequalities within inequalities.'

It is true, of course, that Indigenous people are not a monolith. There are differences in history and culture, as varied as the landscapes they inhabit, but there is also diversity in opinions on various issues as Indigenous peoples deal with new forms of settler colonialism, the challenges and promises of modernity, and the climate crisis. I saw this in the Northern Alaska, where communities are divided on offshore oil drilling, for example. And, as in all sufficiently large human groups, there are always individuals who succumb to the worst aspects of human nature: greed, short-term gain, the desire for power over others.

A key learning for Yuvan emerged from the graduates of the internships: several graduates, many of them women who had not had a previous experience of Nature, came back to say that they wanted to do this work of Nature education full time. The experience shifted their world. He's seen this change even in people in their 40s and 50s. That was the incentive to create Palluyir, a word that means 'all of life.' I am intrigued, because the term erases the false boundary between humans and the rest of Nature. Yuvan tells me that the name comes from a very old couplet of the great Tamil poet Thiruvalluvar from 2,500 years ago. 'To share and protect our biodiversity is the essence of our hoarded wisdom.'

Among Yuvan's inspirations are the writings of Native American scholars like Gregory Cajete (Cajete 2016), Robin Kimmerer (Kimmerer 2015), and Leanne Betasamosake Simpson (Simpson 2014). One of the things he has applied from his learnings is the notion of consent in education. The teacher must have the consent of the student in order for learning to be meaningful and effective. This is a paradigm shift from the common notion of the child as an inferior version of the adult. The child, in this alternative view, is already whole and complete. Therefore, the teacher must explain what they intend to teach and why, and wait for consent. Part of this approach is to 'make consciously central that the educator is a learner … one of the things that is a killer, which has this way of feeding into the industrialized capitalist system: the educator becomes the source of power and hierarchy.' After a Nature walk, teacher and student alike write down their observations. Observing with others, 'interspecies listening,' provides the horizontality of interaction that enables 'the practice of humble wonder.'

I ask Yuvan about his thoughts on my pedagogical framework. He thinks there are some things to learn from it, such as transdisciplinarity, and the use of planetary boundaries. But where there's a lack, he says, is in the direct participation. In his opinion, limiting the learning to the classroom cannot result in a real change. Intellectual discussions are not a substitute for the actual thing, the experience of Nature. Here again, the caution: to think of Nature as inclusive of human, in the 'Palluyir' sense. It is that direct experience, he contends, that can bring about the fundamental change in experience that we need. We have a discussion on what he means by 'direct participation.' He says it's not just about observing other species. It's also interviews, discussions, letter writing, artwork, engaging with local authorities. Even introspection, when properly contextualized within the three pillars of his framework, can be part of it. For example, learning about privilege and its role in climate disruption, and how that behooves us to act. I mention the climate projects that my classes have done, such as designing a heat illness awareness tool for senior citizens in my city. This is also direct participation. But what's missing is the immersion in and the relationship to nonhuman Nature. He's right. This is something I have to think about.

9.3   When People Come and Work Together: Nagraj Adve, Delhi, India

Nagraj Adve is a co-founder of a unique informal network of educators based mostly in India, called Teachers Against the Climate Crisis (TACC n.d.). A non-hierarchical network of primarily college educators, TACC has become – among other things – a way for educators across disciplines to educate themselves and each other about climate change, and take their learnings to the classroom. Lack of the right kind of teacher training is a key roadblock to quality climate education (Kwauk 2020), so this coalition of concerned educators is one way that teachers can – at least partially – fill the gap.

Nagraj is not a college professor, although he has work-related and personal ties with multiple educators. He is a writer and activist, and, before getting into climate change in 2006, had been working for 15 years on issues including democratic rights, equitable urban development, workers' issues, and displacement induced by industrialization – that is, India's development trajectory and its impact on the rights of marginalized people. Climate change seemed to him to connect all these issues, which were often viewed as separate. 'It also occurred to me at the time,' he says, 'that climate change constituted, both implicitly and explicitly, a very strong critique of capitalism in practice and demonstrated that capitalism had effectively failed both people and the planet.'

So, Nagraj and a few friends formed a small group that tried to understand the climate problem and its implications, following which they traveled to Western and Central India (Gujarat, Uttar Pradesh, and Madhya Pradesh) to talk to farmers and see the impacts for themselves. He later went to the Sunderbans (in West Bengal), Tamil Nadu, Karnataka, and Maharashtra to explore the effects of sea level rise and other issues on fishers and other coastal communities. Nagraj has been having regular conversations with urban residents about heat stress, water shortages, and other effects of climate change and how they cope. He has also been delivering talks about climate change at universities and for organizations for the past 16 years.

Education, broadly speaking, should not be limited to students in formal settings, not when the subject is an existential crisis like climate change. For climate change and, indeed, for the polycrisis we face, everyone needs to be adequately informed, from government officials and politicians to those most impacted by the changes. Nagraj's three focus groups are farmers, trade unions, and young people.

'With farmers, it is a question mainly of the impacts they face, as one of the largest social groups whose livelihoods depend directly on the vagaries of Nature, though they may not necessarily attribute it to what is understood as anthropogenic climate change,' he says. India's farmers are mostly smallholders. Many engaged in full-time agriculture are women. Their vulnerability makes reliable climate information a justice issue.

What about trade unions? Workers in most industries tend to be lower-income. They usually have immediate crises to worry about, such as job security, poor working conditions, and the threats to workers' rights over the past 25 years, Nagraj says. In the past 4 or 5 years, the leadership of a handful of trade unions has begun to realize that an energy transition is underway and inevitable, and will affect millions of working people in coal mining, thermal plants, and transport sectors, among others. This has catalyzed greater engagement with climate change. However, 'some unions in the coal sector are resistant to a transition away from coal and uncertain about the expanding renewables sector, in which the wages and working conditions are far more adverse than of permanent coal workers.'

How does Nagraj see climate awareness and concern among the young people with whom he interacts? He says that 15 years ago, he would focus on convincing them that climate change was real and serious. Now, they don't need convincing – it is a (small) part of their curricula in high school; its impacts, especially in the intensity of extreme weather events, are evident; and nowadays information circulates on social media and TV. In addition, some towns have youth climate action groups, such as Fridays for Future ("FridaysForFuture India" n.d.) and others.

> Their responses are three-fold: some of them view (and question) endless growth and elite consumption as being at the roots of climate change and other ecological crises; the issue of air pollution has also come up far more often in recent years. Two, politically, climate change is very much placed within the developed–developing country framework, with the refrain being that industrialized nations are responsible for the problem, and now that 'we' are developing, India is being asked to hold back its progress due to climate change. A third response, which comes up in every single talk or conversation is: what can we do? Their own answer to this question is often individual responses. I often need to nudge them to think of collective measures and their relevance.

Nagraj sees a major challenge in communicating climate change to young people: the need to give them hope. 'There has suddenly been a lot of despair about addressing climate change in time over the last few years. A number of them feel we have only a few years left to save the planet and there is growing anxiety among the youth about this.'

In general, Nagraj has been focusing on three key aspects of the climate crisis:

- Communicating climate and Earth system science in an updated and accessible manner
- Making clear the interconnections of climate change with broader processes such as urbanization, multiple structural inequalities of income,

resources, and wealth, energy use, and the spread of capitalism over the past few decades
* Discussing energy transitions in India and elsewhere

With regard to the first point, Nagraj has recently published a small, accessible, and low-priced book called Global Warming in India: Science, Impacts and Politics (Adve, Nagraj 2022). Published now in several Indian languages, this little book is already making its way across schools, colleges, and organizations in India. A book on climate change written for the general public in India that is not only meticulous in its treatment of the science but also goes beyond the numbers to explore the political and justice aspects of the problem is rare to find in the Indian context. Published by Eklavya, a unique educational NGO that sees education as a means to foster social change, the book is a small treasure, especially the last chapter, 'What we can do.' Including, but going well beyond individual solutions, the book presents the necessity of collective action as well as some practical examples.

I ask Nagraj about approaches to communication that he has found useful. He says that many people, including students, find short videos of climate impacts in the Indian context to be helpful. In his presentations, he asks people about impacts where they live – state, town, or village. Giving time for people to articulate their own thoughts, suggestions, and questions is crucial for deeper engagement. It is also helpful, he says, to use metaphors to convey the significance of aspects of the science or certain numbers, such as emissions data. Many people's eyes glaze over when numbers are presented to them and some numbers are unavoidable. So, making numbers meaningful is important.

One of the ways that Nagraj has helped extend climate change knowledge and information is through TACC, the educators' collective he founded along with others in September 2019. As a non-hierarchical group, TACC has no official structure and no designated spokesperson. When the group speaks to the world at large through public statements, for example, it does so as a collective. It is run by a small, rotating group of volunteers and the group communicates mostly through an email listserv and phone messages on a social media group. It also receives no funding, whether from private entities or from the government.

I joined TACC about six months after its founding, and the discussions and knowledge-sharing have been profoundly transformative for me personally. The group includes educators who work on climate science and policy, energy, agriculture, issues of social marginalization and inequity, economics, humanities, the physical and biological sciences, as well as people simply interested in learning and communicating the crisis more effectively in their classes and contexts ('educator' doesn't only refer to formal education).

The impetus for forming TACC, Nagraj says, was three-fold: the realization that the understanding of climate change, and climate science in particular, among most college teachers was very limited; two, courses were being changed to the detriment of environmental issues; and three, it was a way of reaching out to more students via their teachers.

TACC began with meetings and film shows on climate change in Delhi University. But a few months later, COVID-19 hit, followed by a nationwide lockdown. The group had to shift online. So far, TACC has organized over 30 online talks, including talks by scientists from India and abroad as well as social scientists, economists, and others. TACC is also building an online resource for teachers. In 2022, TACC held an online workshop for school teachers in India, which was enthusiastically received. TACC has also issued public statements on government policy and actions – for example, a public statement critiquing a proposal to water down climate topics in the national middle school curriculum in 2022 gained wide newspaper coverage, ultimately compelling the government to back down, at least temporarily.

TACC faces several constraints, the major one being teaching loads and work commitments of most of the members. 'It is difficult for members to work consistently around something that is outside of their immediate obligations. Also, the structure of research and academic production rests a lot nowadays on funding and projects, making it a challenge to work consistently under the non-funded, non-project framework that TACC follows. A third challenge is that we are the only group of its kind in India; hence, we have no fraternal organizations to discuss issues and challenges with and learn from,' says Nagraj.

Nevertheless, a group like TACC has tremendous potential, according to Nagraj, especially in the area of developing resources for teachers, organizations, and communities, and working with concerned social groups. Reaching small towns is a necessary next step in its evolution.

My own observation is that although TACC activities often fall short of its members' ambitions due to the aforementioned constraints, there are some aspects of Indian – or South Asian – culture that have endured despite modernity that make it more successful than it should be. One such aspect is a certain fluidity about time. In the United States, people tend to carve up the time axis into watertight compartments – work, family, etc., one thing following another. I've noticed that in India there is a far more fluid attitude toward time that includes doing several things at once – childcare and attending a Zoom meeting, for example. But this sense of time also implies porous boundaries between work and non-work – meetings on weekends, for example, or evenings of working days. 'We'll adjust' is a phrase and attitude that I've grown up with in India, and it seems that despite the enormous culture change that modernity and globalization have brought, some remnant of this still exists.

About my own pedagogy, Nagraj feels that the framework is potentially useful, and parts of it adaptable to the Indian context. He suggests that I make visible the importance of addressing the problem. What needs to be done about the problem and what can we do? In the United States, he points out, the current discussion around the Green New Deal, the Infrastructure Investment and

Jobs Act and the Inflation Reduction Act can be brought into the classroom. 'Discussing these enable students to relate the issue to their own (American) political context.' He mentions the importance of climate movements. 'There has also been, in my view, a remarkable political movement around climate change in the US, possibly more than anywhere else in the world, including the Sunrise Movement in which numerous young people have participated, and whose mobilization has forced climate change to become a party political issue.' He points out that the recent acts would likely not have seen the light of day without these political movements.

In fact, I do have students study climate movements – as elaborated in Chapter 7 – and indeed, many students write in their reflections that it makes them feel hopeful to know that there are people acting on the crisis. But Nagraj's comment makes me realize the importance of connecting students with these movements, not just having them look these up. I recently joined the email list for the Sunrise movement, and next semester I plan to invite local youth leaders to talk to my classes. This would open a discussion on a just technological-economic transition, and allow me to more explicitly intro- duce the political dimension, which is currently subsumed under 'transdisci- plinary.' I would have to think of creative ways to introduce this into a physics course without cutting out more content than I currently do, but once we admit the necessity of including these topics, the possibilities open up.

Similarly, Nagraj's emphasis on 'what can we do' resonates. In Chapter 7, I have explored a set of criteria for evaluating a proposed climate solu- tion: real solution or false solution? The classroom discussion on solutions and what real solutions might look like in future worlds (through imagining alternative futures) does appear to be effective to an extent. When community service projects are possible, such as my freshman seminar's presentation of a heat illness awareness infographic to the city health department, this is even more so. But it is not enough. One barrier I have found is cultural – although many of my students are immigrants, I have a sense that America's highly individualistic culture makes it difficult to envisage and take collective action. This is where more community service classroom projects and conversations with youth leaders of climate movements can help shift the consciousness.

One of the strong agreements between Nagraj's approach to communicat- ing climate change and my own framework is the centrality of equity and justice. In his book, Nagraj summarizes four aspects of equity that I think are pedagogically useful: (1) equity between people in terms of power, governance, more equal access to land and water, women's empowerment, development; (2) equity in the capacity for resilience and risk reduction with regard to climatic and related crises; (3) equity between generations ('How can we live sustainably if we so deeply valorize growth and consumption?'); and (4) equity between species, the necessity to discard the anthropocentric view, because life is, after all, a web of relationships with humans as one of 1.7 million known species on our planet.

What a challenge this is, in the face of an increasingly worsening crisis! At the end of his book, Nagraj says: 'But history teaches us that positive social change happens when people come and work together, and therein lies hope for the future.'

9.4 'The Struggle for Climate Healing Is the Struggle for Political Power,' Aahana Ganguly, Sundarbans, India

Some 3 years ago, in the middle of the COVID-19 pandemic, Aahana Ganguly started working with young people in one of the most climate-vulnerable regions of the world: the great mangrove forests of the Sundarbans. Here, the delta of three rivers creates a labyrinth of small islands; the shores of land and isle are fringed with mangroves.

'The Sundarbans is perhaps best known as is the largest stretch of estuarine mangrove forests in the world and the home of the Bengal tiger but it is also inhabited on the Indian side by approximately 5 million people scattered across 54 islands,' says Aahana. 'The other 48 of the 102 islands on the Indian side are reserve forests. My work is with communities who live in the inhabited islands in the active delta.'

Aahana's work in the Sundarbans has a roughly circular trajectory: she spent many formative years as a child in the region; later, while obtaining a Ph.D. in Chemistry at Princeton, she completed a 2-year fellowship program on science, technology, and environmental policy at the Princeton Environmental Institute. Trained in chemistry and materials science, she thinks of herself as a 'climate witness.' She is currently a faculty member at Azim Premji University (APU) in Southern India, many hundreds of kilometers from the Sundarbans. But she spends her breaks and summers in the Sundarbans.

Aahana tells me that the Sundarbans region suffers major climatic impacts that has made the lives of the local people miserable. These include frequent floods, unreliable rainfall, and hotter temperatures, which make subsistence agriculture precarious. Fish are decreasing, amid ecosystem shifts such as the displacement of freshwater fish for brackish water species. All this has consequences for the people of the region. 'The decline and increasing marginalization of livelihoods in the delta region has created the first large-scale climate distress migration in India.'

Many of the people who live on the islands of this region are refugees or descendants of refugees, having been forced to leave their homes in neighboring regions due to flood, famine, and war, beginning with the great Bengal Famine of 1943 during British rule. These refugees, Aahana says, came to the Sundarbans determined to make their homes here. With plentiful fish in the rivers and the fertile land of the delta, they worked hard to build their lives as small farmers and fisherfolk. But now, many residents are having to leave this region due to a complex of reasons that includes climate change. With their history of displacement, Aahana says, 'this feels like a cruel joke.'

Teaching climate change to youth in one of the most marginalized and climate-vulnerable regions of the world is uniquely challenging. Aahana finds this experience very different from her day job, teaching at APU during the academic year. Along with colleagues in physics and biology, she has helped develop a four-course interdisciplinary climate change learning experience for students at APU. This is worth exploring for its own sake, but due to time and space constraints and also because I wish to emphasize the predicaments of the marginalized, we talk mostly about her Sundarbans experience when we meet on Zoom.

Aahana's interest in the environment started in her student days in India in the 2000s. 'Environmental degradation was all around me and impossible to ignore. I also spent a lot of time in my formative years in the Sundarbans. I witnessed the devastating consequence of Cyclone Aila in 2009 and the fragility of the region could not have been clearer to me,' says Aahana.

Human settlements on the islands were constructed after building embankments to protect them from the tides. Now, the sea level outside the embankments is rising while the ground under the settlements is sinking. This bowl-like topography makes residents especially vulnerable to storms. This vulnerability was brought home to Aahana in 2020, when the region was hit by twin disasters: the COVID-19 pandemic and consequent lockdown, and Cyclone Amphan. The cyclone hit in May that year.

It was catastrophic for the millions of inhabitants of the Sundarbans. Homes and cattle were swept away. Embankments were broken and destroyed. Saline water flooded acres of farmland and made them unfit for cultivation. The government said that the region had suffered losses worth thousands of crores [more than 10 billion rupees]. Previous cyclones had resulted in mass migration out of the islands but this time, most migrants had started returning home after losing their jobs in the cities due to the nationwide lockdown.

Aahana began to work with a youth organization in the region in the aftermath of Amphan. This was under the auspices of a three-decade-old trade union of agricultural workers, Paschimbanga Khetmajoor Samity (PBKMS), which had an excellent track record on economic and social issues, from employment guarantee laws, minimum wage, and right to food legislation, to prevention of domestic violence and legal rights of women. While they had not worked on environmental issues before, the increasing ecological fragility of the region was obvious.

'We decided to work together to create a people's pedagogy on the environment and climate change,' Aahana told me. The group has an exponential action strategy, where the learners propagate their learning to others through a 'trainers' training.'

Most of the young people with whom Aahana works are educated in government-run high schools, and some have also attended college. In the early days, many of the youth had a perception that as a consequence of India's rapid development, the people of the Sundarbans were much better off than they had been a few decades ago. While they were concerned about their own livelihoods and employment, they were optimistic and generally uncritical about the direction of development in their areas. This was despite the fact that they knew that thousands of young people were migrating out of the area. They had not connected this migration with failed government policies or climate-related issues. Aahana says that this is understandable, as these young people were not closely involved with their parents' livelihoods of farming, fishing, and forest work. Their school lessons had included environmental science and the fundamentals of the greenhouse effect and global warming. But they had not been taught to connect their formal lessons with the reality of their changing world. Recommended environmental actions were oversimplified to the point of meaninglessness. 'Plastics are bad. Stop using them.' and 'Deforestation is bad. Plant more trees.'

Inspired by the action-competence approach (Jensen and Schnack 1997) in which students develop understanding and skills that enable them to take critical action through collective democratic process, Aahana and her group developed a research-analysis-action cycle as the central feature of the pedagogy.

We started with a video documentation project where the students were asked to interview a person above the age of 60 in their own village about changes in the environment, livelihoods, society and politics. The message from a collection of these oral histories was clear. The consequences of environmental change were making livelihoods unsustainable, migration was driving young people away from the region and the political space for dissent was getting smaller.

We followed up with an app-based survey recording perceptions of changes in the environment, livelihood vulnerabilities and out-migration among people aged 40 and above. This gave the students the opportunity to now analyze larger amounts of data and speak to populations beyond their immediate village communities.

The results of the survey and oral histories formed the starting point for our discussion on climate change in the Sundarbans and we correlated the local observations to the consequences predicted by global experts.

She describes some serendipitous learnings from this experience. A group of students visited a refugee camp on Sagar Island. These refugees were living in deplorable conditions. They had been displaced from the neighboring island of Ghoramara because that island was nearly submerged by rising seas. The desperation of the refugees and the apathy of the government shocked the students, even though they were no strangers to poverty and government inaction. This allowed them to think critically about the much touted 'managed retreat' in the Indian context as a response to climate impacts. It affected them emotionally and viscerally in a manner that no amount of data, statistics, or historical narratives could have done.

Aahana says that 'the alignment of narratives from local observers who have lived and worked in the region for over four decades with the narrative that climate science builds about the fragility of the region has been very convincing and very effective in communicating the gravity of the situation.'

In addition, she says, she has used simple experiments instead of resorting to textbook technical explanations. She uses common household chemicals to demonstrate the thermal properties of carbon dioxide, or ocean acidification, or the thermohaline circulation in oceans.

> I also use a lot of games. Gamification, often, is a very useful way of simplifying a complex social or environmental problem. Addressing agricultural challenges of altered climatic patterns or preparing a resilience plan for an incoming cyclone is confounding because students feel that they lack any practical experience but during a game they are able to strategize effectively and clarify their own conceptions much better than in a straightforward discussion or exchange of information. The Red Cross and Red Crescent Climate Centre has a very diverse repository of games (Red Cross Red Crescent Climate Centre n.d.) that can be easily adapted to various contexts.

Aahana sees the socio-economic challenges of the people to be inseparable from environmental issues. Naturally, the youth are concerned about livelihoods – they know that despite their education, their employment prospects are not much better than those of their parents – upon completing their education, they are likely to end up in low-paying unskilled jobs, or migrate out of state. Therefore, the pedagogy helps connect economic and environmental concerns.

> It helps that I partner with a grassroots organization that has decades of experience sensitizing people to class realities, systemic disenfranchisement, and the nature of capitalist power. They also have their own well-formulated vision of what class struggle means for them and what they want the future to look like. So, I like to think of our approach to action-competence learning and our research-analysis-action framework as being embedded in a broader vision for a future society.

Among the pedagogical challenges are limitations that arise from mainstream education. Students know how to read a thermometer, and, with some help, they can interpret a graph or a map, and analyze data. But they have learned to associate knowledge with book learning.

'They have stopped trusting their own capacity to observe changes in their own environment, changes in the society around them and connect it to what they know from books,' Aahana says. In addition, young people immersed in cable TV and the Internet do not respect the knowledge systems of their ancestors, such as farmers, fisherfolk, and honey collectors.

'That hierarchy of knowledge is something that we are trying to break down. These knowledge systems are also dying out because they are not used anymore.' She gives me an example of bad management by no less than the government forest department, which has banned cutting of mangroves. Yet, forest-dependent community members have been maintaining mangrove trees and promoting their growth through specific cutting techniques. There is ample evidence, she says, that community-managed mangroves on inhabited islands are thriving, while trees in the core forest areas are not. 'A flowering and thriving mangrove tree is also a good source of honey which is a very important non-timber forest product in the Sundarbans economy.'

As in many places in India in the current political climate, there are multiple erasures at work.

Bonobibi, for example, is the guardian goddess of the forests venerated by both Hindus and Muslims of the region. Forest workers (honey collectors and fisherman) traditionally enter the forests after seeking her blessings. Bonobibi and the religion of the forest that she represents is a deep-seated belief system that ties together local ecological knowledge, sustainable practices and the adaptive capacity of the people of the Sundarbans.

But the worship of Bonobibi is declining in the past few years, displaced by mainstream North Indian Hindu festivals that were, until recently, unknown in this region. These festivals have become occasions for the display of political power by the current right-wing Hindu-nationalist ruling party. Such festivals are often staging grounds for communal violence, deepening fissures between religious communities.

There are also challenges specific to women, including the young women in the youth group. Outside the workshop space, Aahana says, society is rigid and patriarchal.

We have a lot of young mothers in our action groups and to enable them, the workshop space often has to accommodate children so women are not excluded. The children sometimes participate in our games or experiments, so I guess in a way we are already doing outreach with the next generation.

Other challenges include the mismanagement and destruction of local resources, through a nexus of corruption and business. One example is shrimp aquaculture, which is undergoing a massive expansion in the islands. Shrimp farmers flood land by making holes in the embankments, which is illegal. During a storm surge, the weakened embankments can break, causing large-scale flooding. 'The whole shrimp farming enterprise is supported by widespread local corruption among bureaucrats, law enforcement and elected officials. It's the same kind of corruption that allows deforestation in the mangrove forests, illegal eco-tourism and hotel construction and, also, prevents regular maintenance of the embankments.'

Aahana expects that the currently benign relationship their group enjoys with the local administration might begin to shift.

> Now we are entering a phase in our project when the youth action groups will move out of the workshop spaces and start holding public meetings talking about the climate emergency in the Sundarbans. Obviously, questions about local corruption that is exacerbating the problem will emerge.

As they make demands to the government for sustainable development of the region, Aahana expects pushback from vested interests and political forces.

> More often than not, henchmen of the ruling party are responsible for illegal shrimp farms, illegal hotel and eco-tourism operations in protected coastal zones and diversion of funds allocated for embankment maintenance. Given the shrinking political space for dissent in West Bengal, I expect significant opposition to our campaigns soon.

Her experience of working with Sundarbans youth in one of the most climate-vulnerable and marginalized places in the world contrasts vividly with her role as a college professor at APU.

> In both places, the University classroom and the workshop space, I am trying to create a transformative pedagogy; a learning experience that will motivate action outside the confines of the classroom.

But Aahana's goal in the two places is very different. At the university, the effort is to provide students with the technical knowhow so that they can pursue careers in climate science or policy. By contrast, in the Sundarbans, her group's aim is to create a youth climate action group, through which the young people of the region will campaign for an ecologically sensitive, sustainable development of the region.

> We want the youth to start community campaigns right now because unlike most of the University students, they are in a region that is already experiencing the consequences of climate change. We want them to convince the communities they live in to start experimenting with adaptation strategies like alternative agricultural methods and think about novel architectural designs for resilient housing. So, first there is a difference in the timescale that I think about as an educator. In the Climate Science course that I teach

at APU, the focus is on the next 50 years, approximately the longest time horizon of business strategy and public policy and, also, the length of a student's career. In the Sundarbans, however, I am trying to create a catalyst for action right now.

Another difference is in what Aahana focuses on with regard to solutions. At APU, there is a much wider lens for looking at solutions, such as changes in the energy system, changes in land use, strengths, and limitations of specific technologies for mitigation.

In the Sundarbans, the solutions span a much narrower space. We talk more about strategies for adaptation and resilience. We focus on housing, agriculture, and other livelihoods and the discussion centers around what makes sense for each island. We are trying to use our pedagogical space to synthesize local strategies.

And, of course, the students are different in terms of their societal privilege.

Our classroom at the university resembles a classroom in the Global North. We have students from many different socio-economic backgrounds but by virtue of the university education that they are receiving now, they will soon be part of the most resource-rich generation India has ever had. They will have access to all the choices that a consumer in the Global North has. So, it becomes important to inculcate a sense of individual responsibility, minimalism, and talk about consumerism and degrowth in the classroom by talking about things like Carbon footprints, for example.

It is very different for the youth in the Sundarbans. 'They are barely hanging on. Talking about a concept like their individual carbon footprints makes no sense there. They are at the receiving end of the effects of the climate crisis for which they are barely responsible, and they have to be given strategies to cope.'
But there are commonalities. One of them is climate fatigue.

It is very hard to steer away from a sense of hopelessness and frustration. Students come to the classroom at the university or the workshop space in the Sundarbans hoping to hear about actions and solutions and have to cope with the complexity of the problem. This sense of disappointment is only exacerbated by the unresponsive and undemocratic political space we now inhabit in India.

The university classroom is an idealized space, a liberal, idealist world in which science informs policy. But in the Sundarbans, 'we encounter a world of real capitalist and state power.'

There, the ground realities make it clear how difficult it is to move away from 'business-as-usual.'

> While we have to acknowledge that science was central in shaping the environmental movement and its demands, the belief that the primary problem with environmental politics is a lack of awareness or denial of scientific knowledge is misguided. We cannot keep believing that if the "masses" truly understood the gravity of the situation, action would follow. We must stop working with the assumption that we can sway politicians and the public towards smart environmental policies by using scientific logic and better policy design in our universities, NGOs and think-tanks. The struggle for climate healing is a struggle for political power.

That is worth reiterating: *the struggle for climate healing is a struggle for political power.*

> As a result, we need to create a working-class climate pedagogy that is not centered only on climate science and decarbonization strategies but also on everyday struggles over access to livelihoods, food, housing, energy, and transportation. This is the kind of people's pedagogy we are trying to develop in the Sundarbans.

Aahana found time to speak with me and communicate via email despite an urgent family situation, so I did not press her for comments on my own pedagogical framework. But as I reflect on the incredibly challenging – and rewarding – work she is doing in the Sundarbans, I see a remarkable blending of the four dimensions of climate change that I describe in this book: the scientific-technological: conveying basic scientific information in an accessible and practical way, the transdisciplinary: centering socio-economic concerns, including justice and power, the epistemological: questioning the dominant paradigm and recognizing the importance of alternative epistemologies, and, of course, the psycho-social action dimension through collective work, training the trainees and exponentiating the learning. Her experience bears out what other researchers have found: that limiting climate education to book-learning does not help students connect the dots between that learning and their experiences, and it does not empower them to act. The focus on empowerment and fighting for justice is central to the pedagogy that Aahana describes. One of my own lessons from speaking with Aahana is that I need to figure out how to enable a collective action spirit in an individualistic culture like the United States. I have been partially successful in some classes, but that experience of community is limited to the space within the classroom walls. One of my takeaways from my conversation with Aahana is the necessity of taking students out into the world, to not only directly experience our biophysical surround, but also to recognize hierarchies of power in the real world. This is logistically challenging where I am. It is also the case – as a student pointed out in response to the story of the Village Women and the Forest – that those who are more likely to suffer from climate impacts and socio-economic injustice, can perhaps be more easily inspired to act than those of us who benefit from our globalized system of destruction. Can the action-competence learning process help break this inertia? While competencies and skills are crucial to deep learning, in mainstream education they are not directly taught with the content – there is a tacit assumption that students will develop these on their own, or that these are unimportant and can be ignored. Integrating competencies and skills into content through meaningful experiential learning would be transformative for education.

Another takeaway is the importance of the collective for educators as well. For most of the time that I have been experimenting with climate pedagogy, I have been working on my own, partly because of early experiences of disapproval or dismissiveness. But educators need coalitions too. Imagine being part of a group at one's institution where you work together for transformative change, in the classroom and beyond. Imagine the possibilities of exponentiating the learning into the university community and the city beyond, and the sense of empowerment it must engender in the learner-teacher. How can we enable this in the different spaces that we occupy?

Aahana's work contributes to the scant literature on educators working within marginalized communities to make meaningful change.

References

Adve, Nagraj. 2022. *Global Warming in India: Science, Impacts, and Politics*. Eklavya. https://eklavyapitara.in/products/global-warming-in-india-science-impacts-and-politics.

Cajete, Gregory. 2016. "Look to the Mountain: Reflections on Indigenous Ecology." In *Applied Ethics*, 6th ed. Abingdon: Routledge.

Center for Teaching, Learning and Mentoring. 2022. "Wheel of Privilege and Power." June 2022. https://kb.wisc.edu/instructional-resources/page.php?id=119380.

"FridaysForFuture India". n.d. FFF India. Accessed June 26, 2023. https://www.fridaysforfuture.in.

Jensen, Bjarne Bruun, and Karsten Schnack. 1997. "The Action Competence Approach in Environmental Education." *Environmental Education Research* 3 (2): 163–78. https://doi.org/10.1080/1350462970030205.

Kimmerer, Robin Wall. 2015. *Braiding Sweetgrass: Indigenous Wisdom, Scientific Knowledge and the Teachings of Plants*. 1st Edition. Minneapolis, MN: Milkweed Editions.

Kwauk, Christina. 2020. "Roadblocks to Quality Education in a Time of Climate Change." *Brookings* (blog). February 25, 2020. https://www.brookings.edu/research/roadblocks-to-quality-education-in-a-time-of-climate-change/.

Pyle, Robert. 2011. *The Thunder Tree: Lessons from an Urban Wildland*. Corvallis, ON: Oregon State University Press. https://osupress.oregonstate.edu/book/thunder-tree.

Red Cross Red Crescent Climate Centre. n.d. "Climate Games–Red Cross Red Crescent Climate Centre." Accessed June 26, 2023. https://www.climatecentre.org/priority_areas/innovation/climate-games/.

Simpson, Leanne Betasamosake. 2014. "Land as Pedagogy: Nishnaabeg Intelligence and Rebellious Transformation." *Decolonization: Indigeneity, Education & Society* 3 (3). https://jps.library.utoronto.ca/index.php/des/article/view/22170.

TACC. n.d. "Teachers Against the Climate Crisis – A Compendium of Sources for Teaching and Learning about the Climate Crisis in India." Accessed June 26, 2023. https://teachersagainstclimatecrisis.wordpress.com/.

10 Reflection-Diffraction
Endings and Beginnings

10.1 Introduction

A scholar writing a book of this kind would be expected to provide, at the closing, a critical self-reflection and a conclusion. How effective is the pedagogical framework I have described? What does it accomplish, and where does it falter or fail? What are the next steps? I begin with such a reflection, but for reasons that I hope will become clearer soon, I find myself compelled to go beyond a critical self-reflection toward a diffractive cogitation, and end, not with a conclusion, but with an opening.

To motivate this, let us examine, first, the term 'reflection' and its connotations. Its roots are in Latin, and its literal meaning is 'to bend back.' In geometrical optics, it refers to the bending back of a light ray when it hits a surface. Reflection in physics is a surface process – light (or heat) from the sun is reflected from snow or ice with relatively little absorption, as we learned in Chapter 4. As a metaphor, reflection implies looking back at something, returning to a past event or process or action, and thinking about it. Implicit in the word is the notion of linear time – time as an infinitesimally thin axis stretching from past through the present into the future. Also implicit is that reflection is a solitary endeavor, reminiscent of Rodin's lonely Thinker. And, as Barad puts it, reflection upholds representation of the world as knowledge, separating knowledge and the world (Barad 2007): 'Representation raised to the nth power does not disrupt the geometry that holds object and subject at a distance as the very condition for knowledge's possibility.'

Reflection and reflective practices have a long history in education. As a metaphor and a practice, reflection has its uses, which is why I have devoted a section to it below. However, any acknowledgment of the inherent relationality of the world – as opposed to Newtonian separateness – makes it apparent that learning and teaching are complex processes, and certainly not doable in isolation. Nor is time as simple as we think!

DOI: 10.4324/9781003294443-10

Metaphors are important. As cognitive linguist George Lakoff puts it, 'there are metaphorical ideas everywhere, and they affect how we act' (Lakoff 2014). Lakoff's concept of frames is also relevant.

> Frames are mental structures that shape the way we see the world. As a result, they shape the goals we seek, the plans we make, the way we act, and what counts as a good or bad outcome of our actions. In politics our frames shape our social policies and the institutions we form to carry out policies. To change our frames is to change all of this. Reframing is social change ...
>
> (George Lakoff 2014a)

What I have called a socio-scientific paradigm (as elaborated in Chapter 4 and elsewhere) is essentially a large-scale complex of frames through which we co-create and make sense of the world. Metaphors, like stories implicit and explicit, help shore up frames and paradigms. Therefore, a paradigm shift must involve – among other things – alternative metaphors.

At the end of Chapter 5, I described the physical phenomenon of diffraction. Diffraction is not *just* a metaphor, however. The term 'metaphor' connotes something that exists in the abstract realm of the mind, hovering over and separated from physical reality. Yet, metaphors are grounded in physical experience ('primitive conceptual metaphors... get their meaning via embodied experience' (Lakoff 2014b)) and have material consequences that can be social, economic, and political. Here is Lakoff again:

> Take the metaphor of Labor as a Resource, where companies seek cheap labor, with workers seen as interchangeable commodities to be purchased for minimum cost in a labor market and working people are hired though the "Human Resource Department." Thus, corporations, to maximize profits, should seek to minimize the "cost" of labor—by cutting pay and benefits, outsourcing, and laying off workers whenever possible. Johnson and I saw enormous social and political consequences arising from abstract thought being characterized metaphorically.

Diffraction is both a metaphor and a methodology, or, perhaps, a transmethodology. Reflection connotes a looking back or a return to something past. Here is Barad (2014), playing with 'returning' as 're-turning,' or 'turning over,' as an earthworm does with soil.

> ... the temporality of re-turning is integral to the phenomenon of diffraction ... Diffraction is not a set pattern, but rather an iterative (re) configuring of patterns of differentiating-entangling. As such, there is no moving beyond, no leaving the 'old' behind. There is no absolute boundary between here-now and there-then. There is nothing that is new; there is nothing that is not new.

I attempt to clarify these ideas in Section 10.3. A caveat, however: my discussion of diffraction, and indeed, the extent to which I have tried to apply it, is incomplete and ongoing, and does not do full justice to the concept. I am still in the process of meaning-making with these ideas.

I end this section with quotes from recent work on applying diffraction to qualitative research – intended as provocations and thought-points to be elaborated and illuminated later in this chapter.

On the practice of coding qualitative data in sociological research contrasted with diffractive analysis (Mazzei 2014):

A diffractive reading of data through multiple theoretical insights moves qualitative analysis away from habitual normative readings (e.g., coding) toward a diffractive reading that spreads thought and meaning in unpredictable and productive emergences.

And from the same paper:

Coding as analysis requires that researchers pull back from the data in a move that concerns itself with the macro, produce broad categories and themes that are plucked from the data to disassemble and reassemble the narrative to adhere to these categories.... we found that a focus on the macro was at some levels predictable and certainly did not produce different knowledge in our study ...

In education research (Bozalek and Zembylas 2017):

In reflexivity, there is a researcher as an independent subject who is actually the locus of reflection, whereas in diffraction there is no such distinction as subjects and objects are always already entangled. Thus, from a diffractive perspective, subjects and objects such as nature and culture are not fixed referents for understanding the other but should be read through one another as entanglements.

For environmental education in particular (Brown, Siegel, and Blom 2020):

A positivist perspective views humans as essentially separate entities who must have knowledge downloaded into them, much like data onto a hard drive. An approach such as Barad's may be an important and game-changing path to participatory, relational and generative knowledge building in environmental education and research, given that we need new and creative ways of relating with others (including non-humans) within environmental education to promote socio-ecological justice.

It is my hope that the next two sections, in which I embark on a reflection followed by a diffractive cogitation (with elements of each in the other), will clarify the ideas expressed in the quotes above.

10.2   Assessing Effectiveness – A Relatively Conventional Discussion

Unlike a clearly defined physical system in a laboratory, with a limited number of independent variables, the system consisting of students and educator is by its very nature complex (see for instance (Forsman, Moll, and Linder 2014)). To test the efficacy of any pedagogical approach by conventional means, we need to control multiple variables – however, in the classroom, these variables are not independent. For example, the interaction between one pair of students is not the same as the interaction between another pair of students (unlike the case of molecules in an ideal gas), and this also applies to teacher-student interactions, even if the teacher tries their best to treat each student on an equal footing. How the social-psychological aspect of learning affects the intellectual and cognitive aspect is a subject of considerable research, for example, Yeager and Walton (2011) and we know now that one affects the other. Pedagogical research often involves showing correlations between student success and some intervention, and then attempting to establish causation through a theory or theories of learning. However, there are multiple reasons why I do not take this approach. One is that in order to effectively teach a subject as challenging as climate change, we need to approach it through a framework or a philosophy, rather than a simple intervention or two, and therefore we are dealing with many more interdependent variables than is usual for a pedagogical experiment in a normal physics classroom. For logistical and ethical reasons, I do not set up a control group. Additionally, my classrooms tend to be small (N varies from 8 to 30 students) so I cannot draw any grand conclusions from my results. Although my work developing this approach was carried out over a period of more than ten years following my early failures, it does not make sense to combine the class N values into one set, because different features of this interdisciplinary framework were developed at different times. For instance, storytelling wasn't central to my approach until around 2017. Nor is it possible to treat my study like a collection of pilot studies because various aspects of my approach developed based on student responses to earlier attempts, and I am constantly iterating; the success or failure of one aspect of this framework is likely to depend on the success or failure of another that might have been developed at a different time – in other words, the situation is neither linear or static. The recent COVID-19 pandemic has added new challenges to teaching in general, and teaching of climate change in particular. The proportion of highly stressed (almost to breaking point) students has gone up, in my anecdotal experience. In the eight years before the pandemic, I can only recall one small class (physics for non-science majors) where students didn't respond intellectually and emotionally to the course material (more below on why I think this might be so). Writing in 2023, I can think of two classes in the past three years where students seemed unable to fully engage in the material, either cognitively or emotionally – this includes the physics component as well as climate change. Although stress levels were up in all cases, these stand out. In one class, three of the eight students had major family crises. In another, larger class of science majors, students seemed so exhausted and burned out that they appeared to have no room to care about anything but getting through the course somehow. And yet, two of my most successful freshman seminars also occurred during this period. These experiences have brought home to me the complexity of the classroom as a learning space emplaced in a larger social context, and the crucial, central role of the psychosocial dimension.

Therefore, my pedagogical role in the classroom is one of 'participant-observer,' that is, I am (inevitably) a part of the system I am studying. A deep engagement with every student, including many hours of interaction outside official meeting times, allows me to sense, observe, and respond to cognitive and emotional changes in the student. While the learning from these interactions cannot be quantified, they are a source of useful qualitative information.

Finally, while the intent of this approach is to provide the conditions for students to experience an epistemic shift (a central concept in transformational learning) (Mezirow, J. and Taylor, E.W. 2009; Sterling, 2011, Boström et al., 2018; Macintyre et al., 2018), this is not possible to measure through conventional means. Such an epistemic shift might take place in a series of stages, and its effects may not be apparent until long after the class is over. The lack of transformative pedagogy in other classrooms could erode the impact of this pedagogy over time. Therefore, I cannot claim that my application of this pedagogical framework results in an epistemic shift.

Students' responses (affective and cognitive) to learning about climate change were collected via discussion questions, tests, homeworks, and exams, as well as anonymous surveys, since 2013 (with the exception of Spring 2014, a sabbatical semester, and the pandemic year 2020). As mentioned, responses were also ascertained anecdotally in nonmeasurable ways through constant and deep engagement with students. Since I do not perform detailed statistical analyses for the reasons elaborated above, any statements I make about the success or failure of my approach are primarily qualitative, descriptive, and tentative. A key purpose of writing this book is to make a case for inter-/transdisciplinarity in teaching climate change in a science classroom through methods inspired by transdisciplinary and transformational learning, by presenting the development of one framework (among many possible), and thereby to invite and stimulate responses, critiques, ideas, and further experimentation.

In Chapter 2, I stated that an effective climate pedagogy:

a Equips the student with a fundamental understanding of the basic science, impacts, and evidence of climate change, including its complex, nonlinear nature, as well as the future projections based on various scenarios – *the scientific-technological dimension*

b   Enables the student to understand societal and ethical implications of climate change (climate justice), leading to intersections with economic, cultural, human rights, and sociological issues; to understand how climate change is related to other major social-ecological problems and to critically examine proposed climate solutions from a climate justice perspective – *the transdisciplinary dimension*

c   Enables the student to see the climate crisis as a symptom of a social-scientific framework or paradigm and to understand and articulate the need for new social-scientific frameworks in order to usefully engage with the crisis – *the epistemological dimension*

d   Inspires students to undergo an epistemic shift (Mezirow, J. and Taylor, E.W. 2009), that enables them to explore their own response to the crisis, as well as their agency, and encourages them engage with social-environmental problems in society – *the psychosocial-action dimension.*

Ultimately, we want learners who are informed and empowered in ways that allow them to participate, collectively and individually, in just and effective action for a better world. At an individual level, this requires an epistemic shift in the learner. At a collective level, the classroom must transition from a 'heap' of disconnected individuals to a system, or a community of collaborative learners. The mainstream education system puts numerous barriers in the way of transformative learning, as discussed in Chapter 2, which is why interested educators must find ways around these barriers. This is what has motivated my pedagogy. But in what way and to what degree can it claim to have fulfilled its intent?

Given the caveats in the previous paragraphs, and based on the past five years (not counting 2020) of data, observations, and engagement with my classes, I can make the following statements with a high degree of confidence.

1   Before integrating climate change into my physics courses, I was already experimenting with learning practices such as inculcating a growth mindset, creating a Natural Critical Learning environment, and using embodied learning to enhance students' interest in and understanding of physics concepts. In their anonymous exit surveys, generally more than 80%, sometimes as much as 95% of students reported a significant increase in their interest in physics. About the same proportion report an increase in physics in my classes in the past few years since I integrated climate change into my courses. Therefore, the introduction of climate change topics has not caused a decline in interest in physics

2   An overwhelming majority of students, pre-2020, typically 80 to 90% per class (with the exceptions noted previously) report a new or renewed interest in climate change. A similar majority of students self-report greater

understanding of the scientific aspects of climate change; this increase in understanding is borne out by homework submissions, tests, and exams. The meta-concepts appear to provide a helpful scaffolding for different aspects of basic climate science, from the greenhouse effect and Earth's energy balance to planetary boundaries and climatic teleconnections – however, they need to be used iteratively throughout the semester, and are therefore most successful in my freshman seminar on Arctic climate change. Post-2020, I have mentioned two occasions where my methods were inadequate. However, a small honors class in introductory physics (N = 8) self-reported scores greater than 7 (on a ten-point scale) on understanding of climate science, and the lowest score was 7 (two students). A majority (77%) of students in a freshman seminar in Fall 2022 also self-reported greater understanding of these matters on a Likert scale, borne out by their performance in assignments and discussions. Thus, it appears that this pedagogy shows promise with regard to the basic science of climate change. Similarly, typically 80–90% of students in my classes across courses understand the basics of climate justice – why climate change is a justice issue – based on answers in home-works and exams, and classroom discussions. They also understand how justice is threaded through all aspects of the climate problem, and stories are especially effective in this regard. However, student work on climate solutions and speculative futurism exercises, while encouraging, indicates that I need to do better in helping them develop critical and ethical thinking skills and nurturing the imagination with regard to climate solutions and futures

3 By the end of the semester, most students demonstrate an increased fluency with transdisciplinarity and a systems approach, as indicated by their comfort using Figure 3.1 for oral explanations and presentations in class. They also show an increased facility for drawing and interpreting concept maps. Based on feedback and open responses to the stories as well as class discussions, students appreciate the use of stories in my pedagogy. They tend to remember the stories even if a couple of weeks (or more) have passed since we discussed the last one. Concept mapping helps them understand how different disciplines are embedded within the stories and the real world. In general, students appreciate the transdisciplinary approach, including opportunities for discussion, embodied learning, working in groups, and working on projects that make a difference. They come to see science as relevant to wider concerns in the world. This is based on end-of-semester surveys, where I ask students to rate their responses to these features of the pedagogy on a 10-point scale, and also solicit open responses

4 In critiquing climate solutions via the 5-question framework described in Chapter 7, students demonstrate at least a beginning-level understanding of the ramifications, ecological, social, and ethical, of purported climate solutions. More work needs to be done to develop and apply this framework

5 Student presentations, as well as short-essay questions on exams, indicate a generally adequate understanding of such ideas as paradigms, paradigm shifts, and alternative epistemologies, and many find them revelatory. However, these concepts are challenging, and several iterations and revisits are needed to clarify them. In my general physics classes, there is usually insufficient time to explore Indigenous ways of knowing as deeply as we do in my freshman seminar on Arctic climate change, although they have a better understanding of what Newtonian physics and the Newtonian paradigm entail, and some idea of paradigm shifts in science (geocentrism to heliocentrism, for example). Therefore, the importance of Indigenous ways of knowing is harder to bring out in those classes, although stories like the Village Women and the Forest and The Scientist and the Elder at least serve to point out to students that such ways of knowing exist. Students in my freshman seminar tend to have a deeper appreciation of Indigenous epistemologies, although we do not have a chance to explore Newtonian physics and the Newtonian paradigm in much detail. Guest lectures by experts, and importantly, by Indigenous activists and scholars are especially helpful. Thus, more work is needed on my part to develop better ways of communicating and demonstrating the importance of paradigms and epistemologies

6 Students appreciate the chance to provide regular, anonymous feedback; they appreciate the fact that their concerns and ideas are taken seriously; some have explicitly mentioned that this is one class in which they feel listened to and where their ideas matter. In end-semester surveys, they also rate very highly the opportunity to decide some classroom policies through consensus, as well as second chances on some assignments and exams and extra help hours for every student. My Student Evaluation scores for 'respect for students' are consistently between 4.5 and 5, with 5 being the maximum possible. All this bears out the importance of building a classroom culture in which *every* student is important, and which is collaborative, builds belonging, maintains high standards, and flattens hierarchies, as I've elaborated in Chapter 2

7 Collaborative, small-group work is especially appreciated by most students. There may be a couple of students in a class of 20 who prefer to work on their own, but especially when creative work or presentations for the class are required of them, students generally report a positive experience. Group work is especially important for the psychosocial action dimension. See also Box 7.1 for student quotes from a freshman seminar

8 Students welcome the chance to talk about how they *feel* about climate change. In a reflective assignment with an upper level physics course for science majors, every student ($N = 12$) agreed that emotions matter, both as barriers to and motivators of climate action. Some ongoing work with a colleague, Dr. Deborah McMakin, a professor of psychology, indicates that the modified Grief and Loss Scale (Wysham, Daphne 2012) is helpful for students to know that they need not get stuck in despair or anger, that

emotions shift, and that it is possible to work through them with others in order to act on climate change. Studying climate justice movements and doing group projects that have real-world relevance are especially helpful for students to feel hopeful about the future. I need to work on developing more projects that have direct relevance to vulnerable people and non-humans – as the freshman seminar project on the Heat Illness Awareness tool demonstrated, these can be very meaningful for students. Yet, there are logistical barriers to doing meaningful projects – some may require institutional support. Meanwhile, how students engage with the climate issue emotionally depends also on other aspects of their academic and social lives – as the COVID-19 pandemic showed us, the wider context also impacts the affective aspects of student learning

Therefore, my tentative conclusion is that these disparate pilot studies indicate that the approach I have described points toward an effective pedagogy of climate change, but does not go far enough. Specific points include the following: One, an epistemic shift (by definition an irreversible change) in the learner cannot be demonstrated within the course of one semester in the absence of a longer-term study, although several students describe how they have changed by the end of the semester as a result of the class (see Box 6.1 for some examples). Two, my sense is that the application of this pedagogy does not fully realize its intent as yet. While the basic scientific-technological dimension is brought out quite well, as is the justice dimension, the epistemological and the psychosocial action dimensions still need some work. Whether and how students who already feel disempowered and weighed down with concerns about finances and family might nevertheless feel empowered to act and follow up with action is an unanswered question. Despite promising signs, student engagement with the 5-question framework for evaluating proposed climate solutions indicates the need for better teaching on critical and ethical thinking, while the speculative futurism exercise reveals the need for nurturing the imagination. The classroom philosophy and dynamics as described in Chapter 2 help nurture each student, and I deem these necessary for the success of any pedagogy.

Before discussing future directions, I embark on a freewheeling diffractive cogitation, which is complementary to the afore-going reflection, and, in my opinion, a crucially important way to think differently about the future, and, indeed, about time itself.

10.3   A Diffractive Cogitation

There is no word in English that means 'to think with others' except for the rather unsatisfactory 'brainstorming,' but consider 'cogitation,' which is a contraction of com (together with) and agitate (to stir up, to impel or move). Cogitation generally is taken to be what a person does in solitude, but its etymology suggests otherwise. Imagine the following scenario. A bunch of us – students, educators, scholars, are sitting on the bank of a pond, with our bare feet in the water. We move our feet, making waves. There's an elephant drinking at one end, and a dragonfly skimming the surface closer to us. Fish glide about underneath. There are rocks protruding from the surface at various points. A soft, unseasonal rain begins to fall. Rather to my surprise, I find that the group around the pond includes my grandmother, and the 15th-century Indian poet and weaver Kabir. The ripples from all of these agitations spread out into and through each other, diffracting beyond and between the rocks, combining to create what physicists call interference patterns; these present, simultaneously, the consequences of both difference and similarity, and so generate a new pattern extended in space and time. The surface of the water is like a living tapestry, shifting, undulating, shimmering, glinting with colors as the sun sets in the West.

So, this is the sense in which I use the term 'cogitate' here, and 'diffractive cogitation' to make it a bit more specific to the purpose. Let me stir up my own experience of learning from the climate as teacher, along with rocks, water, nonhuman fellow-beings, my students, and a number of scholars, fellow educators, and activists, among others. In this moment when I write alone in a room near Delhi, India, none of these are present, except for weather – the local manifestation of climate – currently humid air and the sound of incessant rain – and the nonhuman – a vampirical mosquito hovering. Therefore this co-agitation – this imagined gathering at the pond – is seen through my eyes only, as I ponder, simultaneously alone and in company. What do I see on the surface of the water?

The ripples expand, reach toward each other, superimpose, forming patterns that shift in time and space. We are in relation to all that surrounds us, they seem to suggest. Kyle Whyte tells us that time unfolds due to changes in kinship relationships (as opposed to linear time, which unfolds in discrete units). The central problem of our era is the brokenness of our relationships – between human and human, human and nonhuman – arising from, and feeding back into the violent, destructive, exploitative power hierarchies that drive modern industrial civilization. This brokenness manifests as climate change, mass destruction of wilderness and species extinction, and multiple injustices and oppressions among humans. Complex mutual relationalities among climatic and other biophysical systems and human societies are weakened, or destroyed, and generally unacknowledged by our current socio-political-economic systems. If we form, and are simultaneously formed by our relationships, then who or what 'we' are emerges from this mutuality, this intra-action.

In Whyte's view, kinship exists not only between biologically related people but refers to all relationships of mutual responsibility, guardianship, and interdependence. However, cultures that emphasize individualism to an extreme generally deny interdependence in favor of a delusional independence, pitting the individual against the collective, with the cold war bogeyman of the totalitarian Soviet Union offered as an inaccurate example of the latter. But parts of a system need not be subsumed by the whole, and for any healthy, non-hierarchical (in the sense of power hierarchies) system, the whole and the parts must ensure the well-being of each other. Parts here are not pre-defined, but emerge from the material-discursive context. In such a system, therefore, difference and similarity also arise from the context. The 15th-century Indian poet Kabir, a weaver by profession and of ambiguous religious heritage who can be said to have 'diffracted' Hinduism and Islam through both criticism and praise, also speaks of parts and the whole, of difference and oneness as existing in a kind of dynamical simultaneity. Below is my translation of one of his songs.

You are the gardener and the garden
You are the one who picks the flower...
You are the stick and the scale
You are the one who sits in judgment

It is an ancient notion in certain traditions of Indian philosophy that not only does the whole contain the parts, but the parts, in turn, contain the whole, sometimes expressed through the metaphor of Indra's net, infinite in extent, where the jewel at each node reflects the rest of the net in its entirety. I am reminded of a related idea, that of expansion of consciousness, and consequent blurring of identity, through negation. *Neti, neti,* say ancient Hindu and Buddhist texts: 'not this, not this', with reference to the self, to identity, in paradoxical agreement with another ancient phrase, *tat tvam asi,* 'That you are,' where 'that' refers to all that is around you – the garden, the gardener, the flower. So, also, in the songs of the Sufis, whence Rumi and Bulleh Shah deny their socially constructed identities: *I am not Muslim, I am not from Iran,* says Rumi. *Who knows what I am?* says Bulleh Shah. So, when I am in the United States, I am an aging woman of color, and in India I am an aging woman with caste privilege. But what am I when I am walking through the woods, with no other human around me? Perhaps, then, I am just another animal, an Earthling? How very interesting, how freeing, to be able to co-constitute myself depending on whether my fellows are trees or certain kinds of humans!

Buddhist philosophy speaks of the notion of interdependent arising – that everything depends on everything else for its existence. This is naturally reminiscent of the African concept of Ubuntu, last encountered in Chapter 8 during my conversation with Yovita. It is often expressed as 'I am because we are.' As Michael Onyebuchi Eze says in his book Intellectual History in Contemporary South Africa (Eze 2010):

We create each other and need to sustain this otherness creation. And if we belong to each other, we participate in our creations: we are because you are, and since you are, definitely I am. The "I am" is not a rigid subject, but a dynamic self-constitution dependent on this otherness creation of relation and distance.

This excursion into the question of the whole and the parts is not irrelevant to the classroom. According to some recent research from the rapidly developing new field of collective neuroscience, when people talk to each other, share an experience, tell a story together, their brain waves synchronize – that is, brain waves from one person line up with those from another. Functional magnetic resonance imaging shows crests lining up with crests and troughs with troughs (see Denworth, 2023 and references therein). 'The extent of synchrony indicates the strength of a relationship, with brainwave patterns matching particularly well between close friends, or an effective teacher and their students.' To that last point, a fascinating small study in the United States using EEG with teachers and students (Davidesco et al. 2023) found that 'brain-to-brain synchrony was higher in specific lecture segments associated with questions that students answered correctly. Brain-to-brain synchrony between students and teachers predicted learning outcomes …'

We know very little about collective intelligence since the individualistic Newtonian approach, now spread worldwide through colonialism and globalization, dominates knowledge and knowledge production. Can interbrain synchrony in the classroom enhance both collective and individual intelligence? Is this what is happening neurologically in our brains when we have a shared sense of a great class or a good brainstorming session? Can we correlate interbrain synchrony across race, gender, class to mutually cooperative behavior? At present, to my knowledge, these are open questions.

Interestingly, interbrain synchrony illustrates both difference (each participant is distinct) and similarity (some degree of shared understanding and/or empathy) as well as the dynamism of parts and the whole. Depending on our question, the parts may not always correspond to a single individual – perhaps three people in the group generate a similar response and act as a unit until the process shifts and redistributes and redefines the components of the system.

Relationship building in the classroom (both one-on-one and via group work), accompanied by a flattening of the teacher-student hierarchy, as I've described in Chapter 2, is essential for building trust – the foundation for the kind of classroom culture needed for teaching about the polycrisis. Since I am a part of the teacher-students system of the classroom, I cannot talk about changes in the students (as I did in the preceding reflection) without noting how *I* have had to change as well. As a quiet, generally reserved person who prefers to stay out of the limelight, I've had to become more open, take more risks, and push well beyond my comfort zone. Since I encourage 'physics theater' and other enactments, I have also had to learn to be dramatic when the pedagogical need arises. I have had to develop an openness to sharing my own stories of failure and success, and learn to overcome my reluctance to talk about myself when such a conversation may be of use to students. Engagement with the students is a dynamical process, with its own feedback loops – when students respond well to my enthusiasm for teaching, I experience an increase in that enthusiasm, which further enlivens student interest – not dissimilar to the additive feedback loops in the climate system. I must, at every moment, be sensitive to students' responses so that I can respond as needed. Over the years, my students have taught me to be braver than I could have imagined in my early days of teaching.

Although I have got better at building this kind of classroom culture, every new class warrants new learnings and the integration of new ideas from others. And I am always conscious of the one or two students (sometimes three or four) whom I can't reach, who teach me the limits of my approach and push me to keep trying to do better. Along with this, my sense is that a greater boldness – intellectual and affective – is required on my part, commensurate with the urgency of the climate crisis.

A sense of community and connection in the classroom is also essential to realize the psychosocial action dimension, as I discuss in Chapter 7. This is true within the classroom and also beyond it. Estefania, Nagraj, Karen, and Aahana all mention (see Chapters 8 and 9) the need for hope (hope in its more complex sense, as I've noted in Chapter 7). Climate grief and anxiety among the young can push them (and us) into apathy, despair, and inaction. Thus, a 'hope-infused pedagogy' such as Estefania's is crucial. When this is realized through projects that actually make a difference in the real world, then it can be even more powerful. Community projects further clarify the centrality of justice in the problem of climate change. In her school, Estefania engaged her students with local communities, their problems and needs. Aahana's work in the Sundarbans directly involves young people from a marginalized, extremely climate-vulnerable population. Inequality, injustice, and power hierarchies become obvious and contestable. Estefania's career as an environmental researcher was, in fact, sparked by her observation of the impact of rampant development on local people who were forced to sell their land cheaply. As I ponder their work, I know that the lack of direct engagement with vulnerable communities limits the efficacy of my pedagogical framework in my own context.

The ripple effect that meaningful projects can have beyond the classroom is, of course, a wonderful thing for the students' growth and the community they work with. But there are other kinds of ripple effects, widening circles on the pond of my imagination, made real by the work of Yovita, Nagraj and Karen. Yovita's impact on Canadian national policy regarding sustainability and environmental education for teachers, Nagraj's efforts in India with trade unions, farmers' groups, and through the teachers' collective TACC for teachers to teach themselves, and Karen's involvement with Chicago's teachers' unions as centers of change on the climate issue – all of these are inspiring. There is an exercise I have done with students where we map our ripple effect by drawing a concept map of all our social connections, from family and friends to people we know in school and work and beyond. Although I haven't explored this exercise in great depth, it occurs to me that we educators need to do something like this – sit together, preferably over potluck lunch or tea or coffee, or even around a pond at sunset, and map our social networks, see how they branch out, intersect, connect. My conversations with Karen and Nagraj have sparked a determination to see what unions can do in the face of upcoming climatic, biophysical and social changes. Yovita's hard work on influencing education policy at the provincial and national levels reminds me that one of the things I'd like to do with other teachers – at the potluck – is to examine state and national educational policies on climate change and the polycrisis, and see what difference we can make. TACC, working in India, was able to, at least temporarily, halt the removal of key climate related topics in school syllabi. I wonder if a network like TACC is possible in the very different culture of the United States. Teachers coming together, formally or informally, can do so much!

The way that Yuvan and Aahana have expanded their ripple effects is not only different but also inspiring: teaching their students to teach others, and thus exponentiating their impact. So far, all I've done in this regard is to facilitate students in small groups teaching the rest of the class, which has been quite successful. But this does not go far enough. How might I have my students go out into the world and share their learning meaningfully? How might they teach others from different backgrounds to teach as well? I can see a central role for stories, drama, the use of props, and, perhaps a dialog built with the audience in the fine tradition of street theater. As I write this, I remember the streets of Delhi in the early 1980s, when, as a teen, I witnessed street theater for social change at my university campus.

I want to cogitate a bit on the idea of theater before going on. Feminist theorists have pioneered the notion of gender as performance (dressing and acting in a predefined way), and further, that enacting gender also creates it, so that gender becomes an intra-action between social norms and hierarchies (patriarchy) and the body (for a review of the ideas, see Geerts, Evelien 2016). Barad extends this much further by pointing out the active role of matter – beyond the body and beyond humankind – so that we can see that all our concepts arise from the entanglement of the material and the discursive. Thus, the concept of an electron's position arises out of the intra-action of the electron, the apparatus for detecting its position, and the experimenter – and similarly for the complementary concept of momentum, for which the apparatus is different. Concepts, therefore, do not have meaning independent of their material-social context. Therefore, we can regard concepts as performative, emerging from and given meaning through the intra-action of 'inert' matter, humans and nonhumans in specific contexts.

In my practice of physics theater in the classroom, and its adaptation to climate science basics such as the enactment of the greenhouse effect I have described in Chapter 6, this performative aspect can become literal and visible. I am thinking of an extension of physics theater where the human comes on stage with nonhuman elements – animals, landscape, weather phenomena – and asks: who am I and what is my relationship with these others?

Speaking of the nonhuman, the conventional classroom is not an ideal place for what the polycrisis demands of us, even with the alternative furniture arrangements I described at the end of Chapter 2. As Yuvan points out in Chapter 9, the artificial separation of students and teacher from the world, especially the nonhuman world, is hardly likely to be conducive to developing an embodied appreciation for Nature. (Here, I use Nature in the sense of Palluyir, the Tamil word I learned from Yuvan, inadequately translated as 'biodiversity,' which includes humans as well as other species). At the pond of my imagination where we are gathered, 'we' includes dragonflies, fish, a thirsty elephant, and falling raindrops. Karen's emphasis on Nature immersion, especially for younger children but certainly not limited to them, Sonali taking her students for a walk in the forest in her class about Money (both in Chapter 8) – and, of course, Aahana's work in the great mangrove islands of the Sundarbans (Chapter 9) – all these speak to the need to re-establish and bring to consciousness our relationships with other species. As far as I know, there is as yet no research on interbrain synchrony between humans and other species (although it happens *within* species like mice and bats), but we don't need brain wave alignment to acknowledge that we humans are part of and dependent on Nature. As I've mentioned in Chapter 2, there is some work (Broda et al. 2018) on the importance of the feeling of belonging as an indicator of student success on college campuses, so far limited to human-human interactions. In my physics courses, I have tried to instill a feeling of wonder and belonging in the universe as a whole through the story of the origin of the elements that make up our bodies. We live in a star-blind age – the nightly view of a cosmos rich with stars has accompanied human evolution and deeply influenced various cultural cosmologies until recent times, when bright city lights have effectively drowned out the light of stars. The fact that our constituent elements (with the exception of hydrogen) were formed in the cores of stars, and that we – people, animals, rocks, trees, and planets – exist because massive stars like supernovae tend to explode and scatter these elements into debris clouds from which new suns and planets can arise – can help students, at least in the moment, feel part of something much larger than themselves. This is especially the case when we enact, as a class, the formation of the solar system.

But how to reconnect with – or rather, recognize and be responsible to our connection with – our environs and our planet? That we seem to need this connection for our well-being is a subject of much research in environmental education and human health (Frantz and Mayer 2014; Sudimac, Sale, and Kühn 2022). The term biophilia, coined by Erich Fromm and elaborated and extended by the biologist E. O. Wilson (Wilson 1984), describes the tendency for humans to seek connection with Nature. And yet, in my personal experience, the term doesn't go far enough. As a very shy child, I used to often feel excluded from interactions with other children outside the family-close-friends circle, but growing up in Delhi in close proximity with a number of other species, from a variety of birds and squirrels to packs of street dogs, I became familiar with their worlds and predicaments. Befriending these fellow creatures gave me an immense comfort and pleasure. More recently, some eight years ago, I was in the midst of a personal crisis, and happened to be sitting at a table in an outdoor café. Surrounded by strangers and conscious of a profound misery, I was attempting to eat a croissant between sips of tea. Suddenly, a male sparrow alighted on my table, not two feet away from me. He cocked a bright eye at my croissant and then at me. In the dull grey light of a cloudy afternoon, I saw my tiny image in his little eye. I pushed some pastry flakes in his direction, and we ate companionably until he was full. In those few moments, I was reminded that I belonged to something larger than the circumstances of my present affliction. I could, for the duration of that shared breakfast, step out of my misery and experience a relationship older and, perhaps, more enduring than exclusively human-human interactions – the connection between humans and our wild relatives.

The notion of biophilia as an affinity for Nature can contribute to the perception that humans are separate from the rest of Nature (see the critique of the half-Earth idea in Chapter 7). There is a distancing implicit in the conventional use of the term Nature itself, which allows its appropriation by neoliberal power structures (Bogert et al. 2022) with the introduction of nature-based asset classes and the use of market terms like 'natural capital,' 'nature-based solutions,' and 'natural resources.' How short-sighted and strange this perspective seems when we consider Indigenous cultures' notions of other species as relatives! Collaborative reciprocity, as I've elaborated in Chapter 6, acknowledges complex relationality, societal self-limitations and ecological boundaries, and a balance of power between humans and nonhumans. It also implies a blurring of boundaries, a sharing of spirit between humans and nonhumans. Applying the Ubuntu concept beyond humankind to other species, it is literally true that 'I am because we are.' Parvati Devi (in the story of the village Women and the Forest) knows this and acknowledges this in the collective work of the village women to regenerate their forest. She and her companions are motivated to protect the forest for people *and* the animals.

Note that the human-nonhuman relationships implied in the term 'collaborative reciprocity' are necessarily engendered *in place,* shaped by the particularities of lifeforms, geographies, and histories in that location. The surface of the pond trembles with ripples large and small, but the shape, size, and frequency of the wave pattern in a particular location depend on the local conditions – a jutting rock, or a crow drinking – as well as the large-scale effects. To understand what's happening at the surface of the pond, we must be ready to get our feet wet in this place and that one, and then rise up with the crow into the air to get a bird's eye view. That is, meaningful connections are inevitably nurtured through immersion in the local context. It is difficult to relate deeply and in the long term to global abstractions like 'Nature' if one does not have the localized, visceral repeated experience of relating in place with resident nonhumans. Hence, also, Yuvan's emphasis on a pedagogy that is locally rooted and locally relevant, before it pans out to the global. But is a habitual, everyday experience of connection with the nonhuman possible in an urban environment?

The modern city is, in fact, a good example of the separation of humans from the rest of Nature. Currently, more than 50% of the human population lives in cities. Today's cities are designed to exclude nonhumans, except for plant life that is often non-native. Smooth, manicured lawns and trees in long, straight lines indicate an ethos of control, of Newtonian order. Cities are enormously resource-hungry, especially those in the Global North, but increasingly in the Global South as well, drawing from 'natural resources' all over the world. Perhaps it is not a coincidence that cities are also not conducive to human well-being. To quote from a research highlight in Nature Reviews Neuroscience (Yates 2011): 'City living is associated with a stressful social environment as well as an increased risk of mental illness; of note, the prevalence of mood and anxiety disorders is higher in urban areas than in rural areas and the incidence of schizophrenia is greater in people born and brought up in cities. Longitudinal studies suggest that the effects of the urban environment on mental health disorders are causal ...'

In several student writings over the years, I see a yearning for a deeper connection with other species. These are not necessarily sophisticated analyses of human-Nature relationships, but rather, expressions of what is missing in our cities.

So, for instance, during a speculative futurism exercise, one small student group came up with the following in response to a prompt about a person looking out of a window into a city's nightscape:

> Once upon a time Bambi was getting ready to go to sleep in her busy city. As she was closing the curtain, she looked out as night fell on the city, watching the dim street lights gradually come on. She noted how these street lights had been designed to reduce light pollution so as to not bother the many animals living in the city. These lights were also powered by solar panels that absorb sunlight throughout the day, being an example of the clean power sources the city was running on...

There is a wonderful ambiguity about whether Bambi is human or a deer. This is a city co-habited by many animals, designed so that the animals are not bothered by light pollution. Here is an attempt to integrate modern city functionality with 'many animals living in the city!' It calls to mind urban geographer Jennifer Wolch's conception of a city as a Zoöpolis (Jennifer Wolch 2017), which arranges the cityscape in accordance with the natural landscape and deliberately invites the nonhuman into the city. Of course, plants and animals do exist in cities, but (when they are not domesticated) they are usually in hiding, in the interstices between human-made structures.

Mainstream talk of future cities is generally centered around 'smart cities,' when what we need are 'wise cities.' How to help students connect with the nonhuman in an urban environment?

The situation is complicated by the fact that in New England, where my university is located, centuries of ecosystem destruction and the vanishing of keystone species have led to a rise in tick populations carrying Lyme disease, which is now endemic. Running free through the woods, my 94-year-old neighbor informs me, was once the norm, but it is now a thing of the past in our neighborhood. About 60% of new infectious diseases today are zoonotic – crossing from other animals to humans – and causally related to habitat destruction, including deforestation and plantations (Morand and Lajaunie 2021). The origin of COVID-19 is still unclear in 2023, but habitat destruction continues unabated; as epidemiologists warn (Morens and Fauci 2020), we are entering the age of pandemics.

My small exercise of showing students the bee habitat on campus, humming with wild bees and bright with the flowers of native plants, helps them see the contrast with the neatly landscaped flowerbeds of exotic species that dominate the campus. When Dr. Liebert, our ecologist, points out the much higher degree of connectivity between species in the bee habitat compared to the formal flowerbeds, students start to see beauty differently. I am encouraged by Yuvan's reminder that if we look hard enough, we can find nonhuman life everywhere. I clearly need to look harder and venture beyond campus into our local wilderness.

The idea of belonging – or not – is, of course, concerned with our relationships with the rest of the world, and whether these relationships are strong or tenuous, whole or broken. I am thinking of Sonali's reminder to me in Chapter 8 that we must always ask: *who are we in relation to the world?* If relationships in their dynamic becoming create our shifting, fluid, always contextual identities, then we are not always, everywhere, one thing. Who must we be, what must we become, in which contexts, in order to heal ourselves and mend these relationships? Popular culture tends to promote the idea of identity as fixed, independent of context; similarly, personalities, attributes, and intelligence are assumed to be (mostly) fixed from birth. And yet, from Barad, Dweck, and some older traditions that I've grown up with, 'who we are' is always shifting, contextual, co-created, co-performed with our material and social context. Identity is important in certain contexts – in struggles for justice and equity in America, for example, it matters if one is African-American, and coalition-building for justice is crucial for overturning centuries of oppression. And yet it is also important to know that we needn't be limited by our identities, that identities are not fixed and pre-determined, that we can be, at times, free of them, we can expand the possibilities of who we are, under other contexts and circumstances. Such ambiguity can be emancipatory. Could physics theater, expanded beyond my current practice, help students experience being other than, more than their current conceptions of themselves? Could they even imagine themselves as changemakers for a better future? This would be beyond, even, the epistemic shift of transformative education that I have discussed earlier.

Sonali's account indicates the difficulties of pushing for change in a direction away from what the dominant system dictates. Part of the problem is that in current systems thinking, the observer regards the system from a distance, pre-supposing a hierarchical separation. There is also a separation between emotion and cognition; I think of Descartes, a key builder of the superstructure of Newtonian mechanism, who took his analytical knife and made a cut, severing the universe into primary qualities (atoms and their behavior governed by physical law) and secondary qualities (emotion, for example), a cut that was also fundamentally genderizing. Talking about love for the world, for life, for other species, for Mother Earth – is therefore anathema in academia (Blom, Aguayo, and Carapeto 2020), and a source of much discomfort and embarrassment (especially for academic men, in my experience). Contrast this with Indigenous cultures in which the human considering the nonhuman does so in relationship, with full acknowledgment of the other as a relative, a situation that warrants a mutual caring and responsibility, and, indeed, love. How might such a perspective enable an individual-collective epistemic shift in the classroom?

The ripples on the pond expand, each from their centers to create the crisscrossing, shimmering patterns on the surface. The question arises as to what, in fact, we are really doing. If the crisis is so urgent, why are we sitting around a pond instead of acting? There's no time to be lost!

Indeed, there is no time to be lost. Some would argue that we cannot afford to get our ethico-onto-epistemological acts together first and *then* take action – there's just not enough time. This is true of course, on a linear time axis. And linear time is useful and important, but it is not the only aspect of this complex entity we call time. One might argue that we can think of forming kinships in time, *only if* we have time in the first place, but, to echo Kyle Whyte, what if time flows as a *result* of kinship, of intra-action? As a physicist, I am familiar with the idea that time is relative – the passage of time for two people depends on their relative speeds, and it is also affected by gravity. But I have wondered – and explored this a little in speculative fiction – about whether time might, in fact, be even stranger than we think. What if time is thick? What if, instead of being an infinitesimally thin axis stretching away into infinity, it spreads, bifurcates, has holes in it like Swiss cheese? What if it has structure, like a thick rope, a weave? The poet Kabir at the other end of the pond nods at me. My grandmother is smiling. I remember, when she was alive, how things would arrange themselves around her without her exerting centralizing authority, making to-do lists or other apparent forms of ordering events in time – meals, visits from relatives, which were always ongoing in her generous household, large events such as three-day wedding extravaganzas – all seemed to spontaneously weave themselves into being. As a child, I took all this for granted; it was only in adulthood that I began to marvel at this phenomenon. Looking back, I recall her rare ability to invite trust and confidences simply through her benign presence, and how she could weave relationships through conversation, caretaking, affection, sharing responsibility, and doing things together. When there was an event to organize, people in her vast network would spontaneously take up responsibilities, coming together and coordinating with her and with each other without any central control system. So, aunts, cousins, uncles, nephews, somebody's sister's son-in-law, friends of the family, hangers-on – would accomplish what seemed impossible without a lot of deliberate planning.

I experienced this also as a teen participant in the non-hierarchical, unstructured environmental action group Kalpavriksh in Delhi in the 1980s. We argued vociferously at each meeting, but somehow accomplished a fair bit: pushing the Delhi municipal government to pass an act to protect our local forest, investigating and creating the first comprehensive report on a proposed major dam project, pioneering an approach to conservation with social justice concerns, and producing a regular newsletter. This spontaneous, organic, relationship-based way of acting can accomplish what might seem impossible in a relatively short time. My point in sharing these examples is not to deny the usefulness of a more managerial style of organizing actions, with a central facilitator and plenty of to-do lists – there are contexts in which these would make sense. I am trying to make two points – one, the top-down managerial way is not the *only* way to get things done (and not always the best way, especially when working within a complex system), and two, that, *perhaps*, time flows rather differently in complex, relational networks that are based on mutual trust and caring. This is what I mean by 'thick' time.

If, as I speculate, the thickness of time is an emergent property of complex dynamic relationality, such as when people and materials and nonhumans are working together, then we don't have to succumb to paralyzing fear about the polycrisis. We don't have to think about shifting the paradigm first and then acting – these can happen together, in concert – in fact, the only way we *can* shift the paradigm is through the feedback between experience and cogitation, theory, and practice, like dancing partners getting to know each other's rhythms, and making the dance together.

One observation that I have tried to emphasize in this book is the presence of power hierarchies that, through the destructive, exploitative socioeconomic system they maintain for their benefit, have created the polycrisis that confronts us. These power hierarchies co-opt the crisis to their advantage, as is already happening with the neoliberal takeover of the climate problem. If we are to educate for a world more just, more ecologically harmonious than this one, we have to engage with and foreground the question of power (specifically 'power over,' as opposed to 'power with'; Pansardi and Bindi 2021). In my conversation with Estefania we talked about power hierarchies locally and globally that stand in the way of needed change, and this includes governments indifferent to the well-being of all except the wealthy. Nagraj's study of the impact of India's development trajectory on marginalized communities speaks to this as well. A peculiar blind spot of the spectacle that annual

UN conferences have become is the assumption that governments, through their pledges, will act in good faith. Governments in multiple places around the world, from the US to India, from Turkey to the Philippines, act against their own people all the time, if those people stand between 'natural resources' and corporate profit. Environmental defenders continue to be killed in record numbers every year, many of them Indigenous. While we must deal with and engage with nation-states, corporations, and their leaders because they have power over us, we surely cannot depend on them alone. As Aahana says in Chapter 9, 'the struggle for climate healing is the struggle for political power.'

So, what do we do about power? Does a diffractive cogitation stir up any ideas? Perhaps one important thing we have to do is to be vigilant against tendencies to hegemonize, in ourselves and others. We need to be able to detect how power operates in systems, to see through greenwashing and other forms of deceit, and work to confront it, as Indigenous resistance and youth movements are doing. Power can act overtly, without any masks on, as, for example, in the murders of environmental defenders around the world (Greenfield 2022). It can act covertly, through the maintenance and normalization of the dominant paradigm and value system. It can also co-opt, as in the now-ubiquitous references to climate justice by the elite. This last point makes me wonder: perhaps we should stop talking about justice, and start talking about power. The presence of hegemonizing power is the elephant in the room.

Longer-term work is needed to undermine and flatten systems of power, and here again, we need the mutual interplay of practice and cogitation. I am not, at this moment of writing, cognizant of how thinkers like Barad deal with the question of power through diffractive approaches. What comes to my mind is an image of an ocean wave, a series of parallel crests and troughs, approaching a barrier. The wave beats upon the barrier to no avail, or so it seems. But the repeated attempts have created a little crack in the wall, and every time, a little water gets through. And when it does, we see that the wave spreads out on the other side into a semi-circular diffraction pattern, intra-acting with other waves pushing through other openings, moving past other obstacles. Part of thinking complexly about the systems in which we are embedded is to see both the local and the large-scale, to notice the chinks in the system through which we can push through, and push through, always being changed and making ripples of change, tsunamis of change where none seemed possible before.

10.4   The End and the Beginning

There is no closure; there is only change. Endings are beginnings in disguise.

In my reflection in Section 10.2, I discussed how certain research questions related to my pedagogical framework were answered through student responses and learning. I identified gaps in the pedagogy and its implementation, which suggest ways forward. This is useful to me, and hopefully to my readers as well. (This is why I don't set reflection and diffraction at loggerheads, but, instead, invite them to disturb each other.) However, beginning with predetermined research questions already limits our perception of what we are studying. There is no room for *the questions we didn't know to ask*. If reflection is a practice of stepping back from what we observe, and looking for macro themes that (implicitly) speak to our research framings, as the Mazzei quote in Section 10.1 puts it, 'we found that a focus on the macro was at some levels predictable and certainly did not produce different knowledge in our study ...'

In my reflection, I (for the most part) treated myself as separate from the students in the classroom. From Bozalek and Zembylas: 'In reflexivity, there is a researcher as an independent subject who is actually the locus of reflection, whereas in diffraction there is no such distinction as subjects and objects are always already entangled.'

And, indeed, in my diffractive cogitation with multiple others, I could not separate myself from the subjects of my study and various and unexpected companions. I found, with Mazzei, that a diffractive practice 'spreads thought and meaning in unpredictable and productive emergences.' While there were some common macro-themes in both practices, connections that could not have emerged in the reflection arose during the diffractive cogitation: Kabir and the ancients spoke to Barad, and my grandmother to Kyle Whyte. My fellow educators in Chapters 8 and 9 spoke to me as did the sparrow at my breakfast table eight years ago. Monsoon thunderstorms and physics theater together gave rise to interesting possibilities. I could see myself, my students, other humans, the elephant at the pond, the dragonfly, the sea ice, the polar bear, the rains, as part of a great cosmic drama in which our roles shifted with each act and scene. Without conscious effort, the thought arose that 'we have to stop talking about justice and start talking about power,' at least before we talk about justice. What might this entail? Questions and ideas arose I had never considered before.

From the stirring up of the waters between us – students, scholars, activists, educators, poets, and family – emerge patterns and insights both old and new: the multiple uses of stories, the three transdisciplinary meta-concepts, physics theater, the climate as teacher, the teacher as learner, justice and power, interbrain synchrony across species, complex relationalities, intra-action, Indigenous ways of knowing, thick time. My hope is that this book, despite its flaws and shortcomings, serves as a useful provocation, a wave-front disturbing conventional wisdom, intra-acting with the thoughts, experiences and material reality of its readers so that we can take meaningful climate pedagogy forward, backward, and sideways into the world.

References

Barad, Karen. 2007. *Meeting the Universe Halfway: Quantum Physics and the Entanglement of Matter and Meaning*. Durham, NC: Duke University Press.
———. 2014. "Diffracting Diffraction: Cutting Together-Apart." *Parallax* 20 (3): 168–87. https://doi.org/10.1080/13534645.2014.927623.

Blom, Simone M., Claudio Aguayo, and Teresa Carapeto. 2020. "Where Is the Love in Environmental Education Research? A Diffractive Analysis of Steiner, Ecosomaesthetics and Biophilia." *Australian Journal of Environmental Education* 36 (3): 200–218. https://doi.org/10.1017/aee.2020.24.

Bogert, Jeanne, Jacintha Ellers, Stephan Lewandowsky, Meena Balgopal, and Jeffrey Harvey. 2022. "Reviewing the Relationship between Neoliberal Societies and Nature: Implications of the Industrialized Dominant Social Paradigm for a Sustainable Future." *Ecology and Society* 27 (2). https://doi.org/10.5751/ES-13134-270207.

Boström, Magnus, Erik Andersson, Monika Berg, Karin Gustafsson, Eva Gustavsson, Erik Hysing, Rolf Lidskog, et al. 2018. "Conditions for Transformative Learning for Sustainable Development: A Theoretical Review and Approach." *Sustainability* 10 (12): 4479. https://doi.org/10.3390/su10124479.

Bozalek, Vivienne, and Michalinos Zembylas. 2017. "Diffraction or Reflection? Sketching the Contours of Two Methodologies in Educational Research." *International Journal of Qualitative Studies in Education* 30 (2): 111–27. https://doi.org/10.1080/09518398.2016.1201166.

Broda, Michael, John Yun, Barbara Schneider, David S. Yeager, Gregory M. Walton, and Matthew Diemer. 2018. "Reducing Inequality in Academic Success for Incoming College Students: A Randomized Trial of Growth Mindset and Belonging Interventions." *Journal of Research on Educational Effectiveness* 11 (3): 317–38. https://doi.org/10.1080/19345747.2018.1429037.

Brown, Shae L., Lisa Siegel, and Simone M. Blom. 2020. "Entanglements of Matter and Meaning: The Importance of the Philosophy of Karen Barad for Environmental Education." *Australian Journal of Environmental Education* 36 (3): 219–33. https://doi.org/10.1017/aee.2019.29.

Davidesco, Ido, Emma Laurent, Henry Valk, Tessa West, Catherine Milne, David Poeppel, and Suzanne Dikker. 2023. "The Temporal Dynamics of Brain-to-Brain Synchrony Between Students and Teachers Predict Learning Outcomes." *Psychological Science* 34 (5): 633–43. https://doi.org/10.1177/09567976231163872.

Denworth, Lydia. 2023. "Brain Waves Synchronize When People Interact." *Scientific American*, July 1, 2023. https://doi.org/10.1038/scientificamerican0723-50.

Eze, Michael Onyebuchi. 2010. *Intellectual History in Contemporary South Africa.* 2010th edition. New York, NY: Palgrave Macmillan.

Forsman, Jonas, Rachel Moll, and Cedric Linder. 2014. "Extending the Theoretical Framing for Physics Education Research: An Illustrative Application of Complexity Science." *Physical Review Special Topics-Physics Education Research* 10 (2): 020122.

Frantz, Cynthia McPherson, and F. Stephan Mayer. 2014. "The Importance of Connection to Nature in Assessing Environmental Education Programs." *Studies in Educational Evaluation*, Evaluating Environmental Education, 41 (June): 85–89. https://doi.org/10.1016/j.stueduc.2013.10.001.

Geerts, Evelien. 2016. "Performativity." New Materialism. August 14, 2016. https://newmaterialism.eu/almanac/p/performativity.html.

Greenfield, Patrick. 2022. "More than 1,700 Environmental Activists Murdered in the Past Decade–Report." *The Guardian*, September 28, 2022, sec. Environment. https://www.theguardian.com/environment/2022/sep/29/global-witness-report-1700-activists-murdered-past-decade-aoe.

Lakoff, George. 2014a. *Don't Think of an Elephant! - Know Your Values and Frame the Debate*. White River Junction, Vermont: Chelsea Green Publishing. https://www.chelseagreen.com/product/the-all-new-dont-think-of-an-elephant/.

Lakoff, George. 2014b. "Mapping the Brain's Metaphor Circuitry: Metaphorical Thought in Everyday Reason." *Frontiers in Human Neuroscience* 8. https://www.frontiersin.org/articles/10.3389/fnhum.2014.00958.

Macintyre, Thomas, Heila Lotz-Sisitka, Arjen Wals, Coleen Vogel, and Valentina Tassone. 2018. "Towards Transformative Social Learning on the Path to 1.5 Degrees." *Current Opinion in Environmental Sustainability*, Sustainability governance and transformation 2018, 31 (April): 80–87. https://doi.org/10.1016/j.cosust.2017.12.003.

Mazzei, Lisa A. 2014. "Beyond an Easy Sense: A Diffractive Analysis." *Qualitative Inquiry* 20 (6): 742–46. https://doi.org/10.1177/1077800414530257.

Mezirow, J., and E.W. Taylor, eds. 2009. *Transformative Learning in Practice: Insights from Community, Workplace, and Higher Education*. Hoboken, NJ: Wiley.

Morand, Serge, and Claire Lajaunie. 2021. "Outbreaks of Vector-Borne and Zoonotic Diseases Are Associated With Changes in Forest Cover and Oil Palm Expansion at Global Scale." *Frontiers in Veterinary Science* 8. https://www.frontiersin.org/articles/10.3389/fvets.2021.661063.

Morens, David M., and Anthony S. Fauci. 2020. "Emerging Pandemic Diseases: How We Got to COVID-19." *Cell* 182 (5): 1077–92. https://doi.org/10.1016/j.cell.2020.08.021.

Pansardi, Pamela, and Marianna Bindi. 2021. "The New Concepts of Power? Power-over, Power-to and Power-With." *Journal of Political Power* 14 (1): 51–71. https://doi.org/10.1080/2158379X.2021.1877001.

Sterling, Stephen. 2011. "Transformative Learning and Sustainability: Sketching the Conceptual Ground." *Learning and Teaching in Higher Education* (5):17–33.

Sudimac, Sonja, Vera Sale, and Simone Kühn. 2022. "How Nature Nurtures: Amygdala Activity Decreases as the Result of a One-Hour Walk in Nature." *Molecular Psychiatry* 27 (11): 4446–52. https://doi.org/10.1038/s41380-022-01720-6.

Wilson, Edward O. 1984. *Biophilia*. Cambridge, MA: Harvard University Press. https://doi.org/10.2307/j.ctvk12s6h.

Wolch, Jennifer. 2017. "Zoöpolis." *Verso* (blog). November 16, 2017. https://www.versobooks.com/en-gb/blogs/news/3487-zoopolis.

Wysham, Daphne. 2012. "The Six Stages of Climate Grief." Other Words. September 3, 2012. https://otherwords.org/the_six_stages_of_climate_grief/.

Yates, Darran. 2011. "The Stress of City Life." *Nature Reviews Neuroscience* 12 (8): 430–430. https://doi.org/10.1038/nrn3079.

Yeager, David S., and Gregory M. Walton. 2011. "Social-Psychological Interventions in Education: They're Not Magic." *Review of Educational Research* 81 (2): 267–301. https://doi.org/10.3102/0034654311405999.

Acknowledgments

My first debt of gratitude is to my students over the past fifteen years; it is through their responses, challenges, enthusiasms, and concerns that I have been able to develop the pedagogy described in this book. I am grateful also to Framingham State University's Center for Excellence in Learning, Teaching, Scholarship and Service (CELTSS), to the dean of STEM, Margaret Carroll, and the Chair of the Department of Environment, Society and Sustainability, Larry McKenna, for a course release in Fall 2022 that enabled me to complete part of the work for this book. I acknowledge the support of the office of the Provost over the years, particularly former Provost Linda Vaden-Goad, for enabling me to apply for and obtain a program award from the Association of American Colleges and Universities to create a case study for undergraduate education on climate change in Alaska in 2014–2015. Travel support from my department and from CELTSS enabled me to visit the Alaskan North Slope in 2014, which was seminal to the development of this work. The Framingham State Internal Review Board granted permission for pedagogical studies over the past several semesters, for which I am most grateful.

It is my pleasure to acknowledge also the support and intellectual companionship of a number of colleagues (current and former) across disciplines, especially: Irene Porro, Virginia Rutter, Lisa Eck, Ben Alberti, Rebecca Hawk, Ishara Mills-Henry, Bridgette Sheridan, Deborah McMakin, Lina Rincon, Joe D'Andrea, Brandi Van Roo, Jesse Marcum, Kevin Surprise, Aviva Liebert. Irene Porro's initiative in co-conceiving and running the first interdisciplinary climate pedagogy workshop for high school teachers in Massachusetts in 2017 led to important insights that contributed to the development of this work; I owe her also for a multitude of wonderful transdisciplinary discussions about various aspects of this book. I am indebted to Ben Alberti for introducing me to the work of Karen Barad and also to Joe D'Andrea for rich discussions over the years. My gratitude to Brandi Van Roo, Aviva Liebert, Joe D'Andrea, Ben Alberti, Amy Johnston, Rebecca Hawk, Larry McKenna, and Ellen Zimmerman for guest lectures in my classes and to Judith Otto for useful interdisciplinary conversations and for inviting me to her Global Studies workshops. Much gratitude to Sam Witt for holding climate poetry

226   *Acknowledgments*

workshops in my classes on three occasions. I will be forever indebted to Ishara Mills-Henry and Rebecca Hawk for helping me deepen my understanding of racial and Indigeneity-based oppressions in the United States. Much gratitude to Deborah McMakin for a wonderful collaboration over three years of a pilot project in my classes on the affective impact of climate education. Many thanks to Bridgette Sheridan for discussions on the history of science and for introducing me to the work of Steven Shapin, and to Stefan Papaioannou and Wardell Powell for useful conversations. To Virginia Rutter and Lisa Eck, I owe innumerable pleasurable discussions over tea and meals on a multitude of subjects. Jesse Marcum was the first fellow scientist with whom I could have freewheeling interdisciplinary conversations, from quantum physics to feminist pedagogies. A newer generation of colleagues in the sciences have shown commendable initiative in integrating justice issues in science pedagogy and helping change university culture toward antiracism; I especially appreciate the efforts of Santosha Adhibhatta and Amy Johnston.

For invaluable feedback on my first attempt to write up my work in 2020, I owe Ashok Prasad, Ramaa Vasudevan, Nagraj Adve, Irene Porro, Jeremy Shakun, Henry Huntington, Chirag Dhara, N.D. Hari Dass, James Risbey, Rajesh Kasturirangan, and Shobhit Mahajan. For continued conversations and insights on climate science and climate justice, my gratitude to Chirag Dhara, Henry Huntington, James Risbey and Jeremy Shakun. I am grateful to Christina Kwauk for key insights on the failures of conventional education and for her encouragement of my work. For generously sharing their knowledge with my classes and thereby enriching our understanding of Indigenous epistemologies, I am deeply indebted to Doreen E. Martinez and Kristen Wyman. Special thanks to Henry Huntington for introducing my students and me to the peoples and cultures of the North Slope of Alaska, and putting me in touch with scientists and scholars in Fairbanks and Utqiagvik in 2014. I am deeply grateful to Edna Ahgeak MacLean, Iñupiaq Elder, linguist and educator, for meeting with me during my Alaska trip and giving me invaluable insights on Iñupiaq language, history and culture. I'm also indebted to the scientists in Fairbanks and Utqiagvik, as well as the people of the Department of Wildlife in Utqiagvik for their hospitality and kindness in sharing their insights, especially Todd Sformo, Barbara Tudor, Craig George and Taqulik Hepa. For participation and learnings from an interdisciplinary National Science Foundation project on physics and dance, I am indebted to Folashade Solomon and her team, especially Mariah Steele and Larry Pratt. For continuing to widen my horizons on the climate crisis, my gratitude to Rajeswari Raina and Barbara Harris-White regarding farming and labor, to N. Raghuram on the importance of the nitrogen cycle, and to Rohit Azad and Shouvik Chakraborty on economics and inequality. A special thanks to Chi Huyen Truong and Dan Smyer Yu for giving me the chance to participate in a unique interdisciplinary project with environmental humanities scholars that expanded my intellectual imagination. As a recent member of the scientific steering group of My

Climate Risk (MCR), a lighthouse activity of the World Climate Research Programme that seeks to make climate science meaningful and actionable for local communities, and as facilitator of their working group on education, I am indebted to climate scientists Theodore Shepherd and Regina Rodrigues for their work, leadership, and encouragement. It is a joy to learn from and with my colleagues in the MCR education working group.

I am deeply indebted to the scholars, educators, and activists who agreed to be interviewed for Chapters 8 and 9, thereby sharing their wisdom, experience, and insights with what I hope will be a large audience: Estefania Pihen, Yovita Gwekwerere, Sonali Sathaye, Karen Trine, Nagraj Adve, Yuvan Aves, Aahana Ganguly. Many thanks to Somnath Mukherji and Rajesh Kasturirangan for long and deep conversations over the years on climate justice in India, and our collaboration on climate justice through the as-yet-embryonic People's Climate Network, and also to Somnath for putting me in touch with Parvati Devi, who features prominently in this book. I owe a major debt to AID (Association for India's Development), a volunteer-driven grassroots movement that has allowed me an inside look at the overwhelming challenges faced by marginalized peoples in India as well as their creativity and agency. Similarly, to the scholars, teachers and activists of the India-based Teachers Against the Climate Crisis, I owe multiple cross-disciplinary learnings. Special thanks to the environmental justice action group Kalpavriksh that has deeply influenced the direction of my life and concerns since my teen years. My appreciation and gratitude also to eminent environmental historian Mahesh Rangarajan and environmental justice activist Ashish Kothari, old friends from whom I continue to learn. To Madhusree Mukerjee, heartfelt thanks for much encouragement and rich, transdisciplinary conversations as well as her fierce sense of justice.

I'm grateful for a wonderful long-term relationship with the Center for Science and the Imagination (CSI) at Arizona State University, in particular with Joey Eschrich and Ed Finn; four speculative fiction projects with CSI have been quite literally life-changing for me.

Many, many scholars whom I don't know personally, have, through their work, contributed to my learning, as numerous citations in each chapter indicate. I am deeply grateful to all of them.

My gratitude to Michele Gutlove for contributing the very useful sketch 'Inequality' and for sharing her extraordinary art with the world. Fervent thanks (in no particular order) to Karen Soorian, Chip Nevins, Renu Namjoshi, Madhusree Mukerjee, Michele Gutlove, Rupal Bhatt, Allison Park, Ronna Kabler, Linda Fleger-Berman, K. S. Tsay, Lisa Eck, Virginia Rutter, Irene Porro, Zeynep Gonen, Yumi Park, Pam Schossau, Steve Shervais, Kurt Kremer, Shariann Lewitt, Sarah Smith, Pam Roberts and Lucille Riddle for kindness, friendship, and support. To my old friends Smita Kundu, Kavita Singh, and Shyamoli Chaudhuri, a lifetime debt for a shared love of physics and growing up together through decades.

To my family: my late grandparents, my parents, daughter, niece, siblings and sib-in-laws, as well as aunts, uncles, cousins and family friends, my second-largest debt for making me who I am, and for enabling me to get to the point of writing this book. I am especially grateful to Ashok, Ruchika, and Ramaa for extraordinary support. Many thanks to my aunt Veena for a useful discussion on non-hierarchical organizing. It is a source of much sadness to me that my father did not live to see this book to completion. A person of great integrity and kindness, in the last few years he had become deeply concerned about climate change, especially its impact on the younger generation. This book is dedicated to his memory.

To the nonhumans in my life, in particular, to Bandit's eternal memory, and more recently to three young palm squirrels, Mishku, Motzu, and Momo, for the joyful reminder that we are not alone and our responsibility does not end with the human.

My greatest debt is to Planet Earth, to whom we owe life and breath; an inexhaustible source of wonder and delight, and the best teacher I could ever imagine.

Any shortcomings, errors, or omissions in this work are, of course, my sole responsibility.

Index

234   *Index*

Printed in the United States
by Baker & Taylor Publisher Services